story of the egg—in all its end-
variety—is an almost unbeliev-
of species adaptation and
inal

Stivens clearly knows what he's talking about in this marvellous book
on the evolution of life and the
survival of species. The author
serves up a feast of facts about eggs
and the varieties of reproductive
behavior that have served to protect
them during a thousand million
years."

DAL STIVENS

Illustrated by
Bob Hines

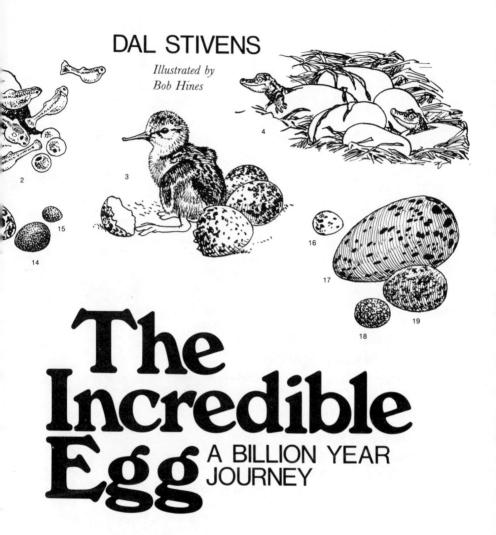

The Incredible Egg
A BILLION YEAR JOURNEY

WEYBRIGHT AND TALLEY
New York

Weybright and Talley
750 Third Avenue
New York, N.Y. 10017

Library of Congress Catalog Card Number: 73-79961
ISBN: 0-679-40049-4
MANUFACTURED IN THE UNITED STATES OF AMERICA

Designed by Bob Antler

CONTENTS

PART ONE ## First Eggs and Fishes

v

PART TWO **Invasion of the Land: Amphibians and Reptiles**

PART THREE **Ultimate Wonder in Birds' Eggs**

Illustrations

First Eggs and Fishes

1

THE STORY IN THE ROCKS

THIS BOOK is about the eggs laid by vertebrates or animals with backbones. ("Animals" is used by most people as a name for warm-blooded creatures with fur, but biologists use it more broadly—for all living creatures.)

Eggs are something most of us take for granted. For a start, we eat them. And we're used to seeing chickens hatching from eggs. Nonetheless, each fertile egg is a miracle.

The beginnings of eggs go back to the dawn of life. The old joke about which came first, the chicken or the egg, is largely meaningless in terms of the story of life on this earth. There were eggs long before the first birds evolved and long before there were vertebrate creatures.

The story of evolution (and of eggs) is told in fossils in the rocks. Early in the last century William Smith, an English surveyor, developed the theory that each stratum or layer of sedimentary rocks had embedded in it fossils that belonged to it exclusively. This pioneer work in geology earned him the name of Strata Smith. Sedimentary rocks include sandstones, shales and limestones and have been built up, one on top of the other, much like a giant layer cake.

Over hundreds of millions of years the sediments of the earth have been built up. Rain, ice and wind have gouged and ground the surface and carried away material to settle on the bottoms of shallow seas, lakes, rivers, streams and ponds or to accumulate in the debris of glaciers.

Sandstone, of course, is made of grains of sandstone cemented together. Shale is solidified silt and clay. Limestone is cemented lime (calcium carbonate), either formed chemically or by the compaction of the skeletons of marine animals. The other water-made rock is conglomerate, which we can loosely describe as cemented gravel. The deposit of sedimentary rocks is immeasurably slow. Erosion, too, is slow but inexorable. Time and time again mountains as high as the Himalayas or the Rockies have been ground down to stony plains. Precambrian deposits in the Rockies are 12,000 feet deep. It has been estimated that the total thickness of the strata of sedimentary rocks is from 60 to 70 miles. Nowhere, of course, are they in such a tidy state as the aforementioned layer cake; earth upheavals have scattered and dispersed them, sometimes burying them deep in the earth. But we can study a bit of the sediment deposited, say, 400 million years ago, in North America, a bit in Europe, another in Asia, and in the strata lies the story of life (and eggs) as it evolved. One reason why the Grand Canyon is so remarkable from the geological view is that in its strata is recorded over 1.5 billion years of earth history.

The scientific study of fossils is a comparatively new science, dating from the publication in 1859 of Charles Darwin's revolutionary book, the *Origin of Species*.

In geologic time there are three great eras of the fossil record: Paleozoic, which means ancient life; Mesozoic, middle life; and Cenozoic, recent life. (These words are formed from the Greek *zōē* = life, and the prefixes *palaiōs* = ancient, *mesos*

= middle and *kainós* = new.) The geologic history before the Paleozoic has been divided into several eras but for convenience' sake is often called Precambrian time.

The eras are divided into sedimentary periods which all bear names. Sometimes they are named after the areas in which they were first studied. Or they are named after their particular characteristics, such as the Carboniferous, which refers to the giant carbon or coal deposits of these rocks.

Cambrian comes from Cambria, the old name for Wales. The next two periods are named after two prehistoric tribes of southern Britain, the Ordovices and Silures, who once lived in the areas where these rocks were first studied. The first was a Celtic tribe which inhabited part of Wales, and the second came from the Welsh borderline. Similar rocks, of course, occur elsewhere but are always called Ordovician and Silurian.

Devonian rocks were first studied in Devonshire, England. Geologists in North America divide the Carboniferous into the Mississippian and the Pennsylvanian.

Permian comes from Perm, a district in Russia. Triassic suggests three and describes the three-fold subdivision of the rocks of this period, studied for the first time in central Europe.

The name Cretaceous is derived from the Latin *creta* = chalk and was first used to describe the cliffs of southern England—remember the white cliffs of Dover?

Tertiary, meaning third, and Quaternary, meaning fourth, refer back to a former classification. The other old names, Primary and Secondary, have been discarded, but Tertiary and Quaternary have been retained for the sake of convenience.

Periods are further divided into epochs. The Quaternary breaks down into the Recent and the Pleistocene, meaning "most recent time." This last epoch covers from one million years ago to 10,000 years ago. In the same way, the Tertiary is broken up into—in descending order—the Pliocene, meaning "more recent," Miocene, "less recent," Oligocene, "slightly recent," Eocene, "dawn of recent times" and Paleocene, which means "more ancient than Eocene."

Few animals which died in the past ever became fossils. To become fossilized, an animal usually had to have hard parts and it also had to be buried quickly before decay destroyed it. And it had to be left undisturbed for millions of years.

Teeth have been preserved with little or no change, but bones generally have undergone changes. Water has dissolved the chemicals from the bones and replaced them with other substances such as silica, lime and iron. (Petrified wood is an instance where silica has replaced the original wood.)

Fossilization of the softer parts of animals is even less frequent. Soft marine animals have died and been quickly covered with mud. In time the mud became shale, and the animals have left behind a delicate "film" of carbon.

Some fossils are molds or casts. The animal has disintegrated, but slowly enough for chemicals to fill the cavity and leave a petrified cast of the original. The tracks of animals have also sometimes been preserved in this way.

Sometimes whole animals have been preserved. During the last Ice Age woolly mammoths were deep-frozen. Animals have also been preserved in peat and in arid regions where the dry air has inhibited decay.

A charming footnote to the story of fossils is that one of the most famous of all fossil hunters was a little girl. When she was only twelve, Mary Anning of Lyme Regis in Dorset, England, won world fame in 1811 with her discovery of the fossilized remains of a giant fish-lizard, the *Ichthyosaurus*. She was the first, moreover, to make a systematic collection of fossil reptiles. It all began when as a four-year-old she helped her father, Richard Anning, gather fossils from the cliffs around their Lyme Regis home. Her father kept a small shop where he sold fossils and shells to people who came to the seaside for holidays. Then when she was still a little girl her father died. Mary decided that she would carry on his work and help to support her mother, brother and sisters. She went almost daily to collect fossils until that wonderful day in 1811 when she stumbled over a fall of rocks and saw some large bones sticking out of the ground. She cleared away some of the rubble, and excitement overcame her, for before her was the skeleton of a giant reptile. She tried chipping with her hammer but soon gave up because she saw that the removal of such a giant fossil was beyond her. She hurried home to get several men to help her. The news spread rapidly, and famous scientists hastened down from London to view for the first time ever the complete skeleton of the extinct *Ichthyosaurus*.

Her find made her famous. During the next few years she found more of the remains of these giant fish-lizards with the paddlelike arms which had swum in the seas that once covered England over 100 million years ago.

When she was twenty-four, Mary won further fame by discovering the first fossil of a plesiosaur, another giant sea reptile. Then, in 1828, she made yet another famous discovery —that of the *Dimorphodon*, one of the first flying reptiles. *Dimorphodon*, which had leathery wings, appeared early in the Jurassic period. In a church at Lyme Regis there's a stained-glass window to the memory of Mary Anning.

The fossil record of animal life on the earth only becomes abundant in Cambrian times, about 600 million years ago. And, surprisingly, these are not of primitive (that is, early) marine invertebrates but of advanced ones such as trilobites and brachiopods (mollusks). This would seem to suggest that animal life, like that of plants, stretches well back into Paleozoic times. Chemical traces of plant life have been discovered from over two billion years ago. (Recent research suggests that the earth is at least 4.5 billion years old.) But the fossil record of animal life is scanty—only a few Precambrian fossils have been uncovered, and it is possible that we may never discover any records of animal life much earlier than those we have found so far. Recent research suggests that the apparently dramatic population explosion of animal life in the Cambrian period could have been due to a sudden increase of plant-produced free oxygen.

Just as the story of life is in the rocks, so, too, is the story of eggs. Indeed, eggs are the essential core of our story because it is by means of eggs that life has evolved and has been passed on from generation to generation.

Along with the story of eggs and their development, the three sections of this book tell the story of parental care—of care, first of all, of the incubating eggs and, later, of the emerging young.

2

THE PROCESS OF EVOLUTION

THE STORY of life on this earth is the story of evolution. Charles Darwin, in his *Origin of Species* (1859), deduced this mechanism of evolutionary change.

Darwin's theory rests on three essential things he observed. The first was that within a single species there was a great degree of variation. This was true of creatures in their natural state and when domesticated. (He was unable to explain why this inherited variation occurred, but since Darwin's time, genetic studies of inheritance—and mutation—have provided an understanding.) His second great observation was of the prodigality of nature, that all creatures tend to increase in geometric ratio. Work since Darwin's day has confirmed nature's astronomical prodigality. A sturgeon may produce upwards of a million eggs a year, a codfish may produce about six million eggs a year, a starfish about 200 million, and an oyster about 500 million. (Not all of these eggs, released in the water, are necessarily fertilized.)

The fecundity of some insects is also impressive. One

African termite lays about 30 million eggs a year; garden worms lay about 100,000.

In theory, the oceans should soon be full of oysters and starfishes. And it has been estimated that a single pair of houseflies and their descendants could cover the earth within a few years to a depth of 50 feet. But it does not happen. Some eggs aren't fertilized and don't hatch. Others are eaten; few of those that do hatch reach sexual maturity.

The third great observation Darwin made was this—that in spite of nature's potentially astronomical prodigality, the numbers of any one species remained fairly constant.

From these facts—the variation within a species, the lavish fecundity of nature and the fairly stable populations— he argued that those creatures which reached sexual maturity and reproduced themselves must in some way have been "fitter" to survive. Their particular variation or difference from their fellows must have been a useful one, and they in turn would pass on these desirable variations to their offspring. Thus, the struggle for survival is not merely with other species but also within a particular species. To quote from one of Darwin's own examples:

> Let us take the case of a wolf, which preys on various animals . . . and let us suppose that the fleetest prey, a deer for instance, had . . . increased in numbers, or that other prey had decreased in numbers, during that season of the year when the wolf was hardest pressed for food. I can under such circumstances see no reason to doubt that the swiftest and slimmest wolves would have the best chance of surviving and so be preserved or selected. . . . I can see no more reason to doubt this, than that man can improve the fleetness of his greyhounds by careful and methodical selection.

Note, too, that the struggle for survival did not mean only "Nature red in tooth and claw," as Tennyson put it. Sometimes a species competed with other species in a particular environment, such as a desert or an icy mountain slope. A succulent, with better ability to store water from infrequent rains, would have an edge on another with less storage ability. A plant with a greater resistance to frost and snow would cover more of the mountain slope than one less well equipped. And, taking an example from birds, a female plover that laid eggs more resembling the terrain would have an advantage over another plover whose eggs were more conspicuous, because those creatures which preyed on eggs would be less likely to see them.

Equally, Darwin stressed that creatures that cooperated with others in the same species and had some degree of social contact had an advantage in the struggle for food and living space.

Every gardener has seen the struggle for existence taking place in his flower and vegetable beds. Darwin cleared a plot of ground three feet long and two feet wide. He watched and counted the weeds that germinated. Three hundred and fifty-seven appeared, but 295 were destroyed, mostly by snails, slugs and birds. No wonder Darwin asked, "Can we doubt that individuals having any advantage, however slight, over others, would have the best chance of surviving and procreating their kind?"

Darwin realized that his phrase "the struggle for existence" might be misunderstood. He wrote in the *Origin of Species*, "I use this term in a large and metaphorical sense including dependence on one another."

Although some of Darwin's ideas have been modified, his brilliant explanation of evolutionary change by natural selection remains central to our thinking today.

Genetics has contributed to our understanding of evolution. It has explained how a variation or combination of variations can occur in a particular creature and how they can be passed on to the next generation. It's one of the saddest accidents that although the basic principles of heredity were discovered by the Austrian monk Gregor Mendel (1822–1884) only six years after the publication of the *Origin of Species*, his work was not known to Darwin and was ignored by almost everyone until the turn of the century.

Modern thinking about evolution has been concerned with combining Darwin's theory of natural selection with discoveries in genetics. This marriage is often referred to as Neo-Darwinism.

The work of Mendel and later researchers has unlocked some of the secrets of inheritance. The way in which an individual person or animal develops is determined by the genes in the fertilized egg. These regulators are made up of DNA (deoxyribonucleic acid). Genes with their coded DNA messages of inheritance are in the cell nucleus where they are arranged in larger thread-like units called *chromosomes*. Each chromosome may have hundreds of *genes*. Because a gene can copy itself, it has been called "the primordial basis of life itself."

All creatures have varying numbers of chromosomes in pairs. The tiny fruit fly (*Drosophila melanogaster*), much used in genetic studies, has eight chromosomes and possibly up to 10,000 genes. A rhesus monkey has 42 chromosomes; a Cebus monkey, 54; cattle, 60; a rat, 42; a cat, 38; and horse, 60. *Icerya* (a scale insect) has 4. Humans normally have 46 chromosomes, and when a cell divides during our normal growth, the two new cells each have 46 chromosomes. But when special reproductive cells are formed—sperm in males and ova in females—then the chromosomes divide by a special process called *meiosis*. A sperm thus has 23 chromosomes, an ovum, 23 chromosomes also. (This process is described in more detail in Chapter 4.) What is important is that genes, those carriers of inheritance, are normally in pairs, but in meiosis these pairs are divided. Only one gene of the pair goes into either the sperm or the egg. These divisions are random; a gamete (male or female germ cell) may have any of the possible combinations of one of each pair of chromosomes. Thus the fruit-fly gamete may have any one of 16 combinations. The formula is 2^x where x equals the number of chromosome pairs. A human gamete can have any one of 2^{23} combinations; that is, 8,388,608!

Any male gamete with any one of these 8,388,608 combinations can fertilize a female gamete with any one of 8,388,608 combinations. Thus more than 70 billion kinds of fertilized eggs are possible. The chance of any human being the same as another is remote, even if they're identical twins. (Identical twins may look alike because a single fertilized egg has divided to form two fertilized eggs, and they may start with identical genetic patterns. They are believed to experience some genetic changes during early development.)

This chance assortment is called independent assortment. The number of possible kinds of gametes is still further increased by what is called *crossing-over* (of the genes); that is, some genes which are normally linked move from one chromosome to another. Variations thus can be immense even in such a simple creature as *Drosophila melanogaster*. Consider that 30 gene changes could produce 1,073,741,824 different combinations—which you have to multiply by 16, the number of chromosome changes in the fruit fly. Consider that the fruit fly has up to 10,000 genes, and we have possibly 10,000, or even 100,000! The number of human combinations is, therefore, astronomical—it has been suggested they could be greater than the number of protons and electrons in the universe!

This combination of independent assortment and crossing-over is called *segregation*.

Genes determine such things as height, weight, skin color. For instance, two or more pairs of genes are responsible for

skin color. And two pairs determine eye color. Most characteristics are decided by the combining of many genes; some are
on different chromosomes so that some cross-over could take
place.

This is the simple mechanism of heredity producing
variety within any species. In addition, brand-new heredity
material called a genetic mutation appears from time to time.
In the fruit fly a mutation appears about once in every
400,000 genes. Most mutations are not helpful. In fact, most
are harmful, and some are lethal. But, nonetheless, even a
small percentage of helpful mutations can greatly assist a
creature in the battle for existence. We shall look at one now.

Most evolutionary changes take place too slowly for us to
observe them. However, the peppered moth (*Biston betularia*) of
northern England provides us with a striking if simple
exception. When it was first observed, before the industrial age
got into its dirty swing, the moth was a pale, speckled
silver-gray which merged well with the lichen-covered bark of
the tree trunks on which it settled. This camouflage which it
had developed gave it some protection from predators such as
birds. Within the population, however, there were also a few
darker moths—produced by a mutant gene. When the factory
chimneys began to belch soot during the nineteenth century,
killing the lichen and blackening the tree trunks, the darker
moths suddenly had the advantage over their now more
conspicuous, lighter-colored fellows. More darker moths survived predatory birds to reproduce themselves. Today they
form most of the peppered-moth population. Indeed, from
1848 to 1948, the percentage of dark-colored moths dramatically rose from under 1 percent to 99 percent!

It's a classic and much-quoted instance of an admittedly
simple evolutionary change. Other comparable evolutionary
changes called *industrial melanism*—Greek *melas* = black—have
been observed in about 80 species of moths in Europe. Most of

these changes have been due to a single mutant gene determining color. Changes of a more complex character in other creatures take considerably longer. For instance the evolution of the modern horse (*Equus*, from *Hyracotherium*—also known as *Eohippus*—a small, many-toed, terrier-size creature) took about 60 million years. The first horses were tropical forest dwellers in North America and Europe and possibly Asia. With four toes on the front feet and three on the rear, they were well adapted to walking on soft forest floors. When the climate grew colder, the forests declined and were replaced by grassy plains. Growing steadily taller, the evolving horses slowly acquired the single hard toe for walking on hard ground, and they grew speedier. They also acquired complicated teeth aptly described by Edwin H. Colbert as "efficient grinding mills" for feeding on hard plant fibers and seeds.

But whether over 100 years or over 60 million years, the principle is the same: Within a particular species, individuals with a distinct advantage over their fellows will have more progeny and eventually displace their less advantaged fellows.

There are still some large pieces missing from the mosaic of evolution. But the basic pieces are in place—natural selection and transmitted variation. Charles Darwin saw a sublime poetry in evolution. "When I view all beings not as special creations, but as the lineal descendants of some few beings which lived long before the first bed of the Cambrian system was deposited, they seem to me to become ennobled."

And he concluded in the *Origin of Species,*

> It is interesting to contemplate a tangled bank, clothed with many plants of many kinds, with birds singing on the bushes, with various insects flitting about, and with worms crawling through the damp earth, and to reflect that these elaborately constructed forms, so different from each other, and dependent upon each other in so complex a manner, have all been produced by laws acting around us. . . . Thus, from the war of

nature, from famine and death, the most exalted object which we are capable of conceiving, namely, the production of the higher animals, directly follows. There is grandeur in this view of life, with its several powers, having been originally breathed by the Creator into a few forms or into one; and that, whilst this planet has gone cycling on according to the fixed law of gravity, from so simple a beginning endless forms most beautiful and most wonderful have been, and are being evolved.

Eggs are truly remarkable things. Consider that hen's egg in your refrigerator. It is probably unfertilized. But if it was newly fertilized, it would contain everything needed to develop into a chicken. Yet you'd be unable with the unassisted eye to distinguish the freshly fertilized germ cell—a tiny white disc—from an infertile one. Or consider the human newly fertilized egg which is extremely tiny but, nonetheless, can develop into a baby. The average germ cell, just large enough to be seen by the unaided eye, is only about .125 millimeters in diameter or about two hundredths of an inch. It weighs about one twenty-millionth of an ounce.

An egg is fertilized when a sperm produced by the male reptile or bird enters a germ cell produced by the female. Sperms are produced in the testes and eggs in ovaries. When a sperm and an ovum (egg) meet, a wonderful process starts.

Here we must say something about that word "egg" which we use in two senses about, say, that common object of food in our refrigerator, the hen's egg. One use of the word is for the whole—the shell and all that it contains. The other is the more restrictive sense of the germ cell or nucleus. You might call it the egg proper and the rest a cradle, a food supply and a private pond to support the growing embryo.

The first man to observe what happens when a sperm and an ovum meet was a 26-year-old German embryologist, Oskar Hertwig. In 1875 he was studying the sex cells of sea urchins.

Hertwig placed an egg on a slide under his microscope and added some seminal fluid. For a time he watched the sperms swimming by means of their whiplike tails round the egg; then he saw one of the sperms penetrate the egg. He observed that the egg soon formed a firm shell which prevented the entrance of other sperms. He noticed that only the head of the sperm entered the egg, and that soon its whiplike tail broke off and disintegrated. The head forced its way further into the yolk. Then it attached itself to the cell nucleus (germ cell) of the egg and linked with it.

Hertwig could not understand what occurred afterwards in the fertilized cell. Indeed, it was not until the early years of this century that biologists made important discoveries about chromosomes and genes in the nuclei of male and female sexual cells. One important pioneer geneticist was the American Thomas Hunt Morgan, with his monumental work on the genes of the fruit fly, *The Genetics of Drosophila*, published in 1925. Chromosomes are tiny, paired, threadlike structures in the nucleus. Most are only about 1/5000th of an inch long. They carry the genes, also paired, which are smaller still. Ten thousand loci (positions) can lie on a chromosome and each locus can have ten alletes (genes).

Oskar Hertwig also did not understand how the sperms and eggs are produced in the testes and ovaries. What happens—as later embryologists discovered—is that special cells form both in the male parent and in the female parent. Let us see how these cells form. For simplicity, we'll look at the development of a male reproductive cell.

The growth of a reproductive male cell (gamete) begins when a cell divides by the process of mitosis to reproduce itself exactly. The two new cells have the same kind and number of chromosomes as in the original parent cell. (Growth of all cells in our bodies is taking place continually by mitosis.) Now follows a more specialized kind of cell division called meiosis

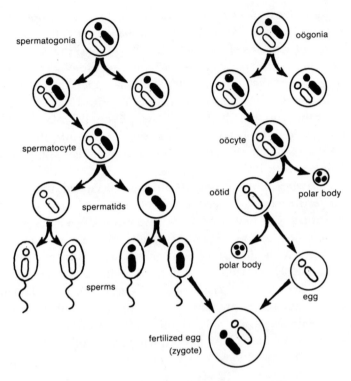

FIG. 1.   The Development of Sperms and Eggs

(or reduction division). The illustration will make this easier to
follow. Meiosis involves two further divisions of each of the two
cells produced by mitosis. Each cell, you will recall, has
exactly the same kind and number of chromosomes as its
original parent cell. (If it is a human cell, for instance, it has
46 chromosomes.) In the first division in meiosis the pairs of
chromosomes in a cell divide in random fashion so that each of
the new cells receives one only of each pair of chromosomes.
Thus each of the two new cells has half the number of
chromosomes in the original cell. If they are new human cells,

each has 23 chromosomes. The cell we have now is called a spermatocyte. The second division follows. The two cells with 23 chromosomes reproduce themselves exactly, as in mitosis. There are now four new cells. The male cells called sperms grow whip-like tails which propel the sperm when it is released, and because the original parent cell divided and reproduced itself in the first instance by mitosis, there are now eight sperms derived from the original single parent cell.

The development of the ovum (egg) is very similar, dividing first by mitosis and then twice by meiosis.

Sperms are produced in billions in the testes. Prodigious amounts are passed from the male to the female at each mating. It is estimated that on average there may be as many as 200,000,000 from a pigeon, 4,000,000,000 from a domestic rooster, 2,500,000,000 from a ram, 4,800,000,000 from a bull, 60,000,000,000 from a boar and 240,000,000 from a man. Nature has a prodigality of which man has taken advantage with artificial insemination. In agricultural breeding centres man has discovered that a certain minimal quantity of semen, much less than that provided by nature, is needed to produce fertilization. Thus one bull from one ejaculation of semen can be made to service many cows.

You will recall that Oskar Hertwig saw extraordinary phenomena under his microscope but could not understand what was happening. Toward the end of the last century the neglected Mendelian laws were rediscovered and Hertwig (now a mature man) and others went back eagerly to their microscopes to peer at the fertilization of the eggs of sea urchins and other simple animals. They saw now that when the egg was fertilized, pairs of chromosomes formed again. Thus when the two cells fuse the fertilized cell once more contains the normal number of chromosomes. They were able to count the chromosomes in the cells of simple animals—and later in more advanced creatures. Thus a fertilized human egg (zygote) has 46 chromosomes. The important thing is that half will have come from the mother and half from the father.

Once an egg is fertilized and receives the required heat to incubate it, growth is rapid. The single fertilized cell divides by mitosis, and within 18 hours in (for instance) a hen's egg, the single fertilized cell, which we can locate as a tiny white disc, has become millions of cells. Within 24 hours the heart begins to take shape and within 44 hours, the heart is beating and blood circulating. (Hens' eggs incubate at a temperature that varies over a few degrees, averaging 101.5° F.)

After about 60 hours a sacklike skin has enveloped half of the yellow yolk. This sac is attached to what will become the stomach of the embryo. At the end of three days the sac has conpletely surrounded the yellow yolk.

At the end of about four and a half days the developing eyes are clearly defined in the embryo's great head, and the limbs are in the bud stage.

When six days have passed, the eyes and limbs are further developed; after nine days digits and toes are in process of forming, the beak is now conspicuous and down appears on the body. After 12 days the chicken is well formed; by the twentieth the chicken has absorbed most of the yellow yolk and begins to breathe through the tiny holes in the shell.

On the twenty-first day when the chicken is due to peck at the shell with its egg tooth, it has taken the last of the yolk into its stomach. It will thereby emerge into the world with a full stomach and will need no food for about 60 hours. Pecking away at the shell, the chicken chips out a narrow strip completely around the shell. It takes about six hours to free itself and emerge wet and clammy.

The development of the embryos of other domestic birds—or wild birds, for that matter—is much the same. Most duck and turkey eggs take 28 days to hatch; muscovy duck eggs take 35 to 37 days, and goose eggs 25 to 30 days.

The domestic hen's egg is an advanced egg—much more complex than the simple sea-urchin egg Hertwig studied, or the eggs of fishes which are the subject of the first part of this book. It is one of those advanced eggs which biologists call the amniotic or land egg (one of nature's greatest "inventions"), but in its first few hours its growth closely resembles all other eggs.

3

CLASSIFICATION: THE SCIENTIFIC NAMING OF ANIMALS

SOMETHING MUST be said here about classification. Scientific names are essential if we are to avoid confusion.

Many creatures have common or popular names, but they aren't accurate enough. Take the common name of "wren" which is applied to many quite unrelated small birds in North America and in other countries. Or take the name "robin" which is one bird in Europe and quite a different bird in North America. If we want to be sure about which wren we are discussing, we must use scientific names—that is, the generic and specific names. Equally, vernacular names such as "wolf" and "tiger" can be confusing because people in different places and countries use them for different creatures. For instance, there are wolves in Europe, North America, South America and India, but they aren't always the same animal.

The classification of animals and plants was pioneered in the eighteenth century by the great Swedish scientist Linnaeus

(Carl von Linné). He chose a binomial (two-name) system which is still maintained today, and he chose Latin because it was the common language of educated men in those days. He had observed that wolves and jackals and domestic dogs all closely resembled each other, so he gave them all the general or genus name of *Canis* (Latin for "dog"). In addition he gave each animal a species name. The domestic dog became *Canis familiaris* (Latin for "of the household"—hence domestic). He called the common European wolf *Canis lupus* (Latin for "wolf"). The Oriental jackal was given the name of *Canis aureus* (Latin for "golden"). The generic name or the name of the genus can loosely be compared to our own last names, and the specific name (or name of the species) to our first names, except that we are stating the last name first.

If we say *Canis lupus* then, it is clear to everyone that we mean the common European wolf. (The American wolf, sometimes called the "gray" or "timber wolf," is considered a variety of *Canis lupus.*)

If we say *Canis lupus chanco*, then everyone knows we are talking about the woolly wolf, which is a subspecies of the common wolf. (Sometimes this third name is necessary. This is discussed later on.) And if you say *Chrysocyon brachyurus*, then people know you mean the maned wolf of South America which belongs to a different genus and species than the common wolf.

And to return to our wren, if we say *Troglodytes troglodytes*, then everyone knows we mean the small bird which is called the "winter wren" in North America. (Sometimes a trinomial [three-name] system is used when a subspecies or variety has been defined. Thus the British wren, closely related to the winter wren, becomes *Troglodytes troglodytes troglodytes*. It's a mouthful, but you could on occasion write it as *T.t. troglodytes.*) In a far-ranging species of birds, those in the middle of the range are true to the definition of the genus and species, but those on the boundaries have acquired so many differences, because of isolation, that they are called a *race* or *subspecies*. Thus two subspecies of the horned owl, *Bubo virginianus*, are the Montana horned owl, *Bubo virginianus occidentalis*, and the Arctic horned owl, *Bubo virginianus subarcticus*.

By the way, Linnaeus sorted out the problem of the European and American robins. He called the former *Erithacus rubecula*, which translates as "a kind of bird inclined to redness." The American robin he named *Turdus migratorius*—"thrush which migrates." The European blackbird belongs to the same genus, *Turdus*, as the American robin but is a different species. Linnaeus called it *Turdus merula*, meaning "thrush blackbird." Thus scientific names are a kind of Esperanto.

A botanist, Mr. Paul Brock, once had a clear demonstration of the practical use of scientific names. He was plant collecting in a remote corner of Corsica and met a native herbalist, an ancient peasant who spoke a dialect that Mr. Brock couldn't understand. But thanks to the use of scientific names, Mr. Brock and the Corsican peasant were perfectly at home over the plants they found.

"*Morisea hypogea?*" * Mr. Brock asked, waving a questioning hand around the landscape. "He led me straight to *Morisea*—which I had come to collect and had so far failed to discover, and would almost certainly have missed altogether but for the native herbalist and the Latin language."

The generic name always has a capital letter. The specific or trivial name does not have a capital. It is usual to put them both in italics. Sometimes throughout this book you may see the name of a genus followed by the abbreviation "spp." This means that there are a number of species within that particular genus. Thus we might say "*Canis* spp."

---

* A member of the sedge family.

The names used are either Latin or latinized names. The latter are derived from classical Greek or from modern surnames and names of places. The name of the genus is a singular noun, and the name of the species is an adjective which must agree with the noun according to the rules of Latin grammar. Thus the emu is *Dromaius novaehollandiae*. The generic name is derived from the Greek *dromaios*, swift-footed, and the specific name is the latinized genitive (possessive) of the country in which the emu is found naturally—Australia, which was once called New Holland. Thus we have the "swift-footed bird of New Holland."

In some reference books you will still see the genus given as *Dromiceius*—which leads us into one of the fascinating bypaths of nomenclature. *Dromiceius* was the name given to the emu in 1816 by the original author, the French naturalist Louis Vieillot, but corrected by him a few pages later on.

Only in recent years have ornithologists reverted to the generic name *Dromaius* that Vieillot intended. You may see in some textbooks *Dromaius* (*"Dromiceius"*) *novaehollandiae*—the discarded name or synonym is put in parentheses.

Another instance is the entry for the veery which may be given as *Catharus* (*"Hylocichla"*) *fuscescens*.

Because *Felis* is feminine, we have *F. lybica* for the Kaffir cat of Africa.

Specific names may, like *"novaehollandiae"* in the case of the emu, indicate the habitat of the creature. Other examples are *africanus* (Africa), *cambricus* (Cambria—Latin for Wales), *gallicus* (Gaul—France), *ibericus* (Iberia—Spain and Portugal) or *nipponicus* (Nippon—Japan). A water plant called *Elodea canadensis* is found in Canada. Or the specific names may describe some characteristic of the creature, or may honor some distinguished naturalist. For example, the specific name of a species of frogs in South America, *Rhinoderma darwinii*, honors Charles Darwin. Other naturalists who have been

honored in this way include Mantell and Philipp Franz von
Siebold. Linnaeus honored the French botanist, Père Magnol
(1638–1715) by naming a genus of trees after him, explaining,
"Magnolia is a tree with very handsome leaves and flowers,
recalling that splendid botanist Père Magnol."

An amusing oddity happened when the name *Smithia
sensitiva* was conferred. Most people took it to be a compliment
paid to Sir James Smith, first president of the Linnaean
Society. However, it was a practical joke, because the name
was coined by a botanist who disliked Smith, whom he
considered conceited and touchy. To get back at Smith, he
created this name for an insignificant weed which, as he put it,
was "a low, mean-looking annual, with sensitive leaflets which
fall when touched."

In strict nomenclature, the name of the author (often
abbreviated) who first named an animal or a plant is added
with the date. Thus the full scientific name of the wolf is *Canis
lupus Linn.* (because Linnaeus first named the creature). I have
dispensed with authors' names in this book.

Naming today is supervised by an international body. In
naming a new species, the generic name is usually chosen from
latinized Greek, or from combinations of Greek words, and the
specific name from Latin. Thus the rainbow lorikeet has a
generic name of *Trichoglossus* and a specific name of *moluccanus*.
Trichoglossus is composed from Greek *trichos* = hair, and Greek
glossa = tongue. Thus "hairy tongue" is a description of the
bristly tongue of these nectar-eating members of the parrot
family. *Moluccanus* means "of the Molucca Islands."

When the coelacanth was rediscovered in 1938 near the
mouth of the Chalumna River, South Africa, Professor J. L. B.
Smith named it *Latimeria chalumnae*. The generic name honored
Miss Courtenay-Latimer, the naturalist who first saw the fish;
the specific name is a description in Latin of the area where it
was discovered.

The binomial system initiated by Linnaeus and the names he used in his book *Species Plantarum* (1753) and the tenth edition of his *Systema Naturae* (1758) are the foundation of our method of naming animals today. The system has, naturally, been greatly expanded and modified since Linnaeus's time. He named a total of 564 birds and 4235 animals of different kinds. To date over a million animals have been described and given names, and 10,000 new ones are added every year. (Between 6000 and 7000 of these are insects; only about six birds are added each year.)

Some names which Linnaeus intended to be accurately descriptive have become meaningless or even misleading because new relationships have been discovered, and this has necessitated some changes. For instance, Linnaeus put the lion with the cats—*Felis leo*—but modern zoologists put it and the other big cats, such as the tiger, cougar, jaguar, etc., in a separate genus, *Panthera*.

Both the generic and the specific names must be unique —that is, they are not applicable to any other creature. The genus and species are at the bottom of the ladder of classification. The basic and smallest unit of classification is the species. It is a distinct population of animals that breed freely with each other but does not breed with other members of other populations. It is also the group or population whose members have the greatest resemblance to each other. Species within a genus can sometimes be crossed—such as breeding a lion with a tiger, and a horse with a donkey. The offspring are usually sterile and, of course, they do not occur in nature.

Above the genus is the family. The various dogs (*Canis*) and foxes (*Vulpes*) have much in common, so Linnaeus grouped them all together in the family Canidae. Equally, there are many similarities between the various cats, large and small, so they are put together in the Felidae.

The various doglike and catlike creatures have much in

common. For instance, both groups have special teeth for
tearing flesh, so they can all be grouped together in an *order*
called Carnivora (from Latin *caro* = flesh and *vorare* = to
devour—that is, flesh-eating). Grouped in the Carnivora are
other flesh-eaters such as the hyenas and civets, bears, weasels,
otters, seals, etc.

Above the order—and still climbing up the ladder—we
have the subclass and class—in the case of the carnivores the
Mammalia (from Latin *mamma* = breast) and Eutheria
(placental mammals). (The other living subclasses in the
Mammalia are the Metatheria [marsupials] and the Protheria
[the egg-laying mammals, the platypus and the spiny anteater.] *)

Modern classification tries to show the way animals
evolved and their degree of relationship to each other. Thus
we can see that *Canis aureus*, the Oriental Jackal, is a
"brother," as it were, to the Common European Wolf, *Canis
lupus*. And, again, the European Wild Cat, *Felis sylvestris* ("of
the woods"), is a "sister" of *Felis chaus*, the Jungle Cat of Africa
and Asia. But both of these are only "cousins," so to speak, of
the *Panthera* genus (lion, tiger, jaguar, etc.), and the wolf is a
"second cousin" of the cats, large and small.

So far so good, but higher classifications of orders, classes,
subphyla and phyla don't show the evolutionary relationships
so clearly. In the various classes and phyla with equal ranking
are animals which followed each other up the evolutionary
tree—sharks, bony fishes, amphibians, reptiles, birds and
mammals. At the top of the animal kingdom are about 20
major groups called phyla, plural of Latin *phylum*, meaning a
race or tribe. There are a number of these such as Protozoa
(single-celled creatures—amoeba, foraminifera, radiolaria,
etc.), Porifera (sponges), Mollusca (clams, oysters, mussels,

---

* These words are made up of *theria* = animal, from the Greek *therion* = beast,
and prefixes, *eu-* = good, *meta-* = after, and *pro-* = before or early. Thus, "first
mammals," "later mammals" and "most advanced animals."

etc.), Arthropoda (joint-footed insects and crustaceans), Annelida (leeches, earthworms). The one we are concerned with is the phylum Chordata (from Greek *chordē* = string). Within it are some seemingly unlikely related animals such as vertebrates and sea squirts and lancelets which have in common a *notochord*, a hollow dorsal nerve cord, and gill slits. In vertebrate embryos the internal backbone forms around a flexible rod, the notochord. Mammals have gill slits only in the embryo stage.

Within the phylum Chordata the most important subphylum is that of the Vertebrata—fishes, amphibians, reptiles, birds and mammals. They differ from the other divisions in the subphylum such as the lancelets and sea squirts in having a skull which contains a brain, and in having a skeleton of cartilage or bone.

The first vertebrates were fishes—the earliest were jawless fishes called *"Agnatha"* (from the Greek *A* = without and *gnathos* = jaws). From these primitive fishes (whose descendants, the lampreys, still swim in our seas) evolved all the other fishes and later the amphibians, reptiles, birds and mammals.

The subphylum Vertebrata has two great divisions or superclasses—Pisces (fishes) and Tetrapoda (from Greek *tetras* = four and *pous* = foot). These, in turn, divide into a number of classes.

Subphylum	Superclass	Class
Vertebrata	Pisces	Agnatha, jawless vertebrates
		Placodermi, primitive jawed vertebrates
		Chondrichthyes, sharks
		Osteichthyes, bony fishes
		Acanthodii, spiny fishes
	Tetrapoda	Amphibia, amphibians
		Reptilia, reptiles
		Aves, birds
		Mammalia, mammals

All this is fairly clear, even if we have had no training in zoology, because the Pisces are obviously different from the Tetrapoda. Equally, within the classes, you can see the differerce between sharks and bony fishes, which include most of the fishes in our waters today. Incidentally, of the five classes of fish, two—the Placodermi (armored fishes) and Acanthodii (spiny fishes)—are extinct. We can also readily see that amphibians, reptiles, birds and mammals are all unlike.

Below the classes are the various orders.

Within these ascending divisions of species, genus, family, order and class, there are intermediate higher and lower grades such as subclass, suborder, etc. and superfamily. For instance, the superfamily Canoidea of dogs, bears, raccoons and mustelids contains:

(a) Family Canidae   dogs, wolves, foxes.
(b) Family Ursidae   bears.
(c) Family Procyonidae   raccoons, coatis, kinkajous, pandas.
(d) Family Mustelidae   weasels, mink, otter, badgers, wolverines, skunks.

To return to our first example of *Canis lupus*, we could set out its full classification like this:

Phylum	Chordata
Subphylum	Vertebrata
Class	Mammalia
Subclass	Eutheria
Order	Carnivora
Suborder	Fissipeda
	(from Latin *fissus* = cleft, and *pes* = foot—that is, with cleft feet)
Superfamily	Canoidea
Family	Canidae
Genus	*Canis*
Species	*lupus*

As you can see from this detailed example, the gradation from the top is into increasingly narrow categories—rather like an inverted pyramid. This can be shown another way if we set out in simplified form in the one table the classification of the domestic dog, the wolf, the European wildcat and the lion. This table will make clear their relationship to each other and to other creatures:

Animal Kingdom (all the animal phyla)

Phylum	Chordata (sea squirts, lancelets, vertebrates)			
Subphylum	Vertebrata (birds, reptiles, amphibia, fishes)			
Class	Mammalia (placental mammals, marsupials, monotremes)			
Subclass	Eutheria (placental mammals)			
Order	Carnivora (Canidae, Felidae, bears, seals, etc.)			
Family	Felidae		Canidae	
Genus	*Felis*	*Panthera*	*Canis*	*Canis*
Species	*sylvestris*	*leo* *familiaris*	*lupus*	
	wildcat	lion domestic dog	wolf	

And we can classify man in this way:

Kingdom Animalia
Phylum Chordata
Subphylum Vertebrata
Class Mammalia
Subclass Eutheria: modern mammals
Order Primates: lemurs, tarsiers, monkeys, apes, men
Suborder Anthropoidea: monkeys, apes, men
Superfamily Hominoidea: apes and men
Family Hominidae: man
Genus *Homo*
Species *sapiens*

An important point is that all groups of any one kind have about the same weight or importance. Thus all families grouped in any one order should have about the same degree of difference from each other. And, equally, each order should differ from the others by roughly the same degree of difference.

Those of you who study Latin will probably have noticed that the name of a genus is always singular, as in *Canis* or *Felis*, whereas categories above the genus, such as family, order, class, etc., are always plural, such as in the example above of man, or in the order Sirenia (sea cows), family Bovidae (cattle group) or family Equidae (horses). The name for the family is always based on the name for the typical genus in the group. Thus, in man, which is unusual for having only one species, the family name is Hominidae. The termination "-inae" indicates a subfamily.

Names higher than those of families, such as those for orders, classes and phyla, don't have uniform endings, although you'll find there's considerable conformity.

You will also find that some experts don't always agree on the precise classification in the higher categories of some animals. This is particularly true of extinct animals, which, naturally, aren't as easy to study as living ones; it's often a question of interpretation of sparse evidence. Only time and the discovery of more limbs and branches in the evolutionary tree will resolve some of the debate. Right now, there's not even general agreement on how many classes of fishes there are—many say four; the latest scholarly evaluation is five (used here), making a distinct class of the Acanthodii (spiny fishes) which was earlier lumped in with the Placodermi (armored fishes). Both are extinct. Again, some experts say there are 27 living orders of birds; others 29. Another expert, equally famous, asserts there are 32. There is disagreement, too, concerning the lower categories. The only stable unit of classification is the species. Above and below it is subjective, a matter of interpretation.

Among zoologists are people whom their colleagues amusingly call "lumpers" and "splitters." Lumpers try to group as many creatures together as possible in a genus, order, class, etc. Splitters try to find and emphasize differences.

You could moralize here that most of us belong to either one or the other of these categories; some of us try to simplify—others are concerned with detail. Lumping or splitting isn't important. But classification is. It's indispensable.

From Aristotle's day, man has tried to identify and classify animals. It's part of his need to know and understand his world. I find it fascinating that a tribe of primitive people in New Guinea had 137 special names for birds which scientists later classified as 138 species. The tribe had confused only one species with another.

A footnote to this chapter is that you can have a lot of fun looking up names in a good dictionary of biological terms, and when you visit Italy or Greece you may find your zoological knowledge useful. A zoologist friend, who confesses to being like Shakespeare in possessing little Latin and less Greek, says he can read much of the inscriptions on early Italian tombs knowing only nouns and adjectives.

4

THE FIRST VERTEBRATES

I N 1774 a strange creature was discovered in the seas off
Cornwall. It was about two inches long and semi-
transparent and looked rather like a flattened cigar. It had no
jaws, no paired fins, no skeleton and no eyes. It was sent to the
great naturalist Peter Simon Pallas, who decided it was a kind
of snail. The creature was given the name "Amphioxus,"
which means "sharp at both ends." For nearly 100 years
Amphioxus was studied and dissected and argued over by
zoologists. Some claimed it was a fish; others doubted if it was
one. Today the question has been resolved. Amphioxus, the
little sea lancelet found in all shallow tropical seas, is not a
fish. It is a very primitive member of the phylum Chordata
(animals with notochords or backbones). Amphioxus has no
vertebrae but has instead a notochord or "back string,"
forming an internal support for the creature. This firm but
flexible slim rod of gristly material enclosed in a tough sheath
is the forerunner of the backbone or spinal column in the
higher-developed fishes—and in ourselves. Above the noto-

chord, Amphioxus has a dorsal (back) nerve cord, but the little sea lancelet has no real head or brain and no sense organs except for some spots of pigment which are probably sensitive to light. It does have excellent gills, used not merely to extract oxygen from the water but to strain food from the debris on the ocean floor where it lives.

It may sound a dull creature but it is of great scientific interest, because when we look at Amphioxus we are looking back some 500 million years. As the famous American paleontologist Edwin H. Colbert points out in *Evolution of Vertebrates*, it is probable that the little sea lancelet has changed little since early Paleozoic or even pre-Paleozoic times. "When we look at a lamprey we get a partial glimpse of the ancient vertebrates that lived half a billion years ago."

Thus the unobtrusive little sea lancelet, which spends most of its time burying itself in the sand, is of abiding interest—a kind of "living fossil." And, to link us more closely to our common ancestry, every vertebrate embryo, including that of man, starts life with a notochord and only later acquires a backbone. Biologists, in fact, often use the story of the development of the embryo of the primitive Amphioxus as a simple introduction to embryology.

Courtship among the sea lancelets is simple, and so is the act of reproduction. Males and females gather together. Males discharge sperm and females discharge eggs through their gill slits into the water, where fertilization takes place. The amount of yolk is small—very much less than in most vertebrate eggs. The fertilized egg divides into two cells, into four cells, into eight cells, until there are some hundreds arranged rather like a mulberry or raspberry. (Biologists, indeed, call the cell cluster a *morula*, which is Latin for mulberry.) The center of the morula dissolves and fills with liquid, and a hollow ball of cells (the *blastula*) develops. Then one side of the blastula begins to cave inward, almost as though the ball of cells was a soft, hollow, rubber ball being pushed in on one side by an invisible finger. The sides of the ball fold to produce a cuplike shape—a sac with an inner and outer layer of cells. Such an early embryo is called a *gastrula*—a "little stomach." The inside of the cup will ultimately become the cavity of the digestive tract of the adult lancelet. The inner layer of cells is called the endoderm (inner skin), and the outer layer of cells is called the ectoderm (outer skin). Between these two layers develops a third layer, known as the mesoderm (middle skin). These layers gradually develop further to form all the structures of the adult lancelet's body. The ectoderm, as you've probably guessed, forms the epidermis or skin and the nervous system; the endoderm develops into the respiratory and digestive tract; and the mesoderm develops into the muscles, the reproductive organs and circulatory system. The larva takes about three months to metamorphose into an adult lancelet.

On February 16, 1894, the great German naturalist Ernst Haeckel, celebrating his sixtieth birthday, offered his guests a delicacy—open-faced sandwiches covered with lancelets served like sardines. He wrote later in his autobiography that this was a dish he would never forget. The reason was that 28 years before this birthday party Ernst Haeckel had developed a fascinating theory which later research has largely supported. The theory was that you could read the evolutionary history of any living creature by studying its embryonic development. The embryo of the lancelet traveled the evolutionary path of its ancestors, from a single-cell organism to a gastrular polyp, to an echinoderm ("prickly skin"), and to a chordate—that is, an animal in this instance with a notochord. These are the same stages through which all vertebrate embryos, including human embryos, pass in the first few days. So Ernst Haeckel did well to serve his guests with lancelet sandwiches!

The lancelet, as we have seen, is a primitive chordate and on a lower rung of the evolutionary ladder than the creatures with vertebrae—sharks, bony fishes, amphibia, reptiles, birds and mammals. Thus it gives us a partial glimpse of what the first chordate animals were like.

Other primitive chordates in our seas include the acorn worms and the sea squirts (which look rather like sponges). A sea squirt begins life as a free-swimming larva and looks rather like a tadpole. It has a notochord and a dorsal nerve cord just as a lancelet has. After a time it abandons swimming, attaches itself to a rock and loses its tail, notochord and its nerve cord.

We shall probably never learn exactly what the first ancestral chordates were like. They were almost certainly small, simple creatures with skeletons much too soft to be fossilized. Zoologists think they probably developed from what may at first sight appear most improbable ancestors—early echinoderms. Modern echinoderms include starfishes, sea urchins, sea lilies and the like.

As Dr. A. S. Romer puts it in *Man and the Vertebrates*, "No one would believe that man or any other vertebrate descended from a starfish!" But, as he points out, the evidence from the development of larvae of the echinoderms strongly suggests that long ago, in the dawn of the world, there existed a very simple creature possessing some of the features of the larvae of the starfish or the acorn worm. From it developed, in the course of evolution, the echinoderms and also the first chordates and, finally, the true vertebrates.

We have a tantalizing glimpse of these first vertebrates in some fossil fish scales found in freshwater sediments in Colorado which date back some 500 million years. Looked at under a microscope, a bony structure shows under the scales. There's also a solitary fossil called Jamoytius from about 400 million years ago, a small, fishlike creature with a torpedo-shaped body. Like the lancelet, it is jawless and has a notochord.

Creatures with backbones, either of cartilage or bone, had many advantages over those without in the struggle to survive. Their strong, flexible skeletons inside their bodies gave them greater freedom of movement than that possessed by the invertebrates, whose shells are external skeletons. (A fish which uses its body to thrust against the water can move much faster than, say, a crab.) Further, the backbone protected the delicate nerve cord which carried messages from the brain. Vertebrates could (and would) develop in the course of evolution a well-developed brain in the front of the body where, despite its apparently vulnerable position, it is protected by a strong bony skull. And, most important, an internal skeleton didn't have the inbuilt limitation of an external skeleton. An external skeleton cannot become very large without becoming weaker. With internal skeletons the way was open for the development of large, mobile creatures.

Some experts think the first vertebrates developed in freshwater; others think that they originated in the seas. The complex arguments for each belief need not concern us here, but with the coming of the first vertebrates, the stage was set for the evolution of the first fishes. They almost certainly evolved first in freshwater and not in the sea, as you might suppose. We can only guess why this should be so. They may have evolved in freshwater seeking peace from fierce sea creatures which preyed on them. Or they may have found rich sources of food.

What is a fish—or, as zoologists say, a true fish? First of all, a true fish is a cold-blooded, water-dwelling animal with a backbone and a skeleton. (We have to be careful of that adjective "cold-blooded." Much of the time a fish's blood is warm. "Cold-blooded" means it adapts its temperature to its surroundings. Biologists prefer to call a fish *poikilothermic*.)

A fish takes in oxygen from the water by means of gills. It has a slimy and usually scaly skin, paired fins and moves about with thrusts of its muscular tail.

By the way, we use the word "fish" rather loosely. Jellyfishes, starfishes and shellfish are not true fishes because, among other differences, they have no backbones. Nor are lobsters, prawns, sponges and corals.

Most fishes lay eggs which are fertilized by the male after they have been laid. Some, however, bear their young alive; their eggs are fertilized inside the females by the males. (You have probably seen tropical live-bearing fish in aquariums.)

Fishes have successfully colonized the seas, lakes and rivers, from the equator to the poles. There are at least 20,000 species.

In the next chapter we'll look at some of the most ancient fishes—the lampreys and hagfishes—and their eggs.

5

ANCIENT LAMPREYS

PEERING DOWN into the clear water of the stream you might easily mistake those long, silvery, arrowing shapes for eels. Their popular name is, indeed, lamprey eels, but they are quite unrelated to eels. If you get a good look at one of the spawning sea lampreys, *Petromyzon marinus,* you'll notice that although the body is scaleless like the eel's, unlike an eel it has a single nostril on the top of its head between the two eyes, that it lacks paired fins on its sides and has a peculiar round, jawless mouth full of sharp teeth.

The lampreys, shimmering below us in the shallow water and undulating upstream, have come from the Atlantic Ocean to mate and die, much as do Pacific salmon (*Oncorhynchus* spp.), exhausted by their labors. In their spring invasion of rivers and streams on both sides of the Atlantic, the two-to-three-foot-long lampreys seek shallow water to spawn. They can even climb falls, provided they're not too high. They do so by attaching their sucking mouths to rocks and pulling their bodies upward. In their mating migration they may travel several hundred miles, not eating at all, so that their bodies continually shrink as they live on the energy stored in their body tissues.

When a male and female lamprey have found a suitable place in a gravel stretch in a shallow backwater, the male excavates a rough nest by carrying away stones held in his sucker, as in Figure 2. (The generic name *Petromyzon* means stone-sucker.) In this fashion he makes a shallow depression about six inches deep and two to three feet in diameter. Here the female will later lay up to 200,000 small, soft, jelly-coated eggs. Fertilization takes place when the female attaches herself to a large stone and the male attaches himself to her head. They coil vigorously while the female's eggs and the male's milt are violently expelled. Their movements stir up the sand, which swirls into the nest and covers the eggs. Then, exhausted and often mutilated by their reproductive labors, they separate and drift listlessly with the current, and die soon afterward.

Although much modified in the course of evolution, lampreys—and hagfishes—belong to the very primitive and ancient class of fishes mentioned earlier, the jawless Agnatha. They are the descendants of the first fishes, which were also jawless. They lack bones—have no vertebrae or ribs; body support is provided by a basketlike framework of cartilage or fibrous material. They have an unsegmented notochord, much like the one possessed by lancelets. They have other primitive characteristics: They lack paired fins, a sympathetic nervous system and a spleen. We find creatures resembling them in the fossils of the distant Ordovician period—all jawless. From that time, nearly 500 million years ago, the only jawless fishes to survive have been the lampreys and hagfishes. Their eggs are, therefore, among the most ancient laid in the waters of our world.

The eggs of the sea lamprey, after being deposited in the

FIG. 2. *Sea lampreys using their sucker-disc mouths to move pebbles and stones to construct their nests.*

nest, take about two weeks to hatch. Several days after hatching, the blind, toothless, wormlike larvae leave the nest and seek silted stretches where they burrow into the sand or mud. They do have tiny eyes but, as yet, these eyes don't function. Called ammocoetes or prides, the lamprey larvae emerge at night to feed on small animals which they suck in through their disclike mouths. The prides pass at least three years in the streams in this larval state. Then when they are from four to six inches long, they start their metamorphosis or transformation to adult lampreys. The rudimentary eyes grow larger and function; the single nostril moves from the front of the head to a position between the eyes; and the mouth changes and acquires 112 to 125 horny, rasping teeth. The lampreys grow silvery and begin their migration to the Atlantic.

In the oceans they turn to a semiparasitic way of life. They attach themselves to fish with their suction-cup mouths, rasp away scales and skin with their toothed discs and tongues and suck the blood of their victims. They secrete an anticlotting substance to keep the victim's blood flowing. These vampirelike attacks alone rarely result in the death of the victims. But the lampreys may weaken them so seriously that they are killed and eaten by other fish. Or they may die from infections. But lampreys probably don't cause undue damage in the seas. Though they prey on fish, they in turn are preyed upon. Since their numbers apparently remain fairly constant, despite the fact that the female lamprey lays up to 200,000 eggs to ensure that one of the young will reach sexual maturity and return to spawn, the death rate is evidently high. Among many fishes the only safety is in numbers. No care is taken of the young as is usual among birds and mammals. From the moment they hatch, the young must fend for themselves.

When the lampreys made their way into the Great Lakes, they had few natural enemies and rapidly caused a disastrous decline in the yearly catch of lake trout and other fish by commercial fishermen. Forty years ago the yearly catch was 11 million tons. Within three decades it declined to almost nothing. Today the United States Bureau of Fisheries is carrying out a widespread research-and-control program in an attempt to restore the fisheries.

Twelve to twenty months after they become parasites, the lampreys reach sexual maturity. Their teeth become blunt, their digestive systems degenerate, they cease to eat and they start the migration back to freshwater where their life cycle started. They do not necessarily return to the very same waters however.

We have looked at the more spectacular of the lampreys, the sea lamprey (*Petromyzon marinus*). Altogether there are seven genera and over thirty species living in the temperate regions of both hemispheres. Some are restricted to freshwater. Some, too, are not parasites. All have a life cycle similar in

most respects to the sea lamprey. The major difference is that
the nonparasitic ones do not feed at all once they become
adults. Hence they do not grow larger than the larvae they
were before they were transformed. Drawing on their reserves,
they live just long enough to spawn and die.

Courtship among the freshwater river lampreys or lam-
perns (*Lampetra* spp.) of North America (*L. ayresi*) and Europe
(*L. fluviatilis*) resembles a game of blindman's buff or musical
chairs. Their eyes are poor and the lampreys must find each
other mainly by touch and smell. First, in the spring, these
nonparasitic lampreys swim upstream. From time to time they
stop and attach themselves to rocks with their suckerlike
mouths. There they hang, swaying in the current. Some
lampreys do not stay long attached to a rock but restlessly
move from stone to stone. These are males seeking mates.
Eventually instead of a stone they find another lamprey and
attach themselves to it. Should this second lamprey be a male,
they detach themselves and the two drift downstream and soon
separate. But if the second lamprey is a female, she retains her
hold. Thus the two sexes find each other. Later the pair swim
together to a sandy stretch where they mate, winding their
bodies around each other, much as the sea lampreys do.

(A passing thought is that lampreys were regarded as a
delicacy in Europe during the Middle Ages. King John, who
reluctantly signed the Magna Carta, was said to have died of a
surfeit of lampreys. They were sold in New England fish
markets as late as 1850. In the Scandinavian and Baltic
countries, they are still in demand as delicacies—so much so
that some governments have legislated to prevent overfishing.)

So far as we know, Australian waters have two species of
lampreys only, both parasitic—the short-headed lamprey,
Mordacia mordax, and the pouched lamprey, *Geotria australis*.
The first inhabits east-coast rivers; the other lives in western
waters; both mingle in Tasmanian rivers. The short-headed

lamprey grows to 18 inches; the pouched lamprey grows to 2 feet in Western Australia. Little is known of their breeding habits.

Hagfishes superficially resemble lampreys. Like lampreys they have a single nostril, but the snout carries six fleshy filaments (barbels). They have nonsucking mouths with rasplike teeth. They have poor eyes and find their victims by sense of smell. Confined to salt water, some 20 species inhabit the Atlantic and Pacific oceans. Hags do not grow as long as lampreys. The longest is the common Atlantic hagfish (*Myxine glutinosa*), which reaches about two feet, six inches.

From time to time when professional fishermen tip their catches of haddock, hake or cod on deck, they discover some are literally "bags of bones." A hagfish has bored its way into the victim and eaten most of the flesh, leaving the skin intact.

A female lays up to 30 large eggs. While most fish eggs are less than one-fourth inch in diameter, those of the hagfish are about one inch long in a horny shell with a tassel on each end. Unlike lampreys, hagfishes have no spawning period and lay large, yolk-rich eggs throughout the year. There's no larval stage, so the young look like miniature adults.

Not only are hagfishes of great scientific interest in allowing us to board a time machine and move back to the time when the first fishes were evolving, but their primitive hearts make them invaluable laboratory animals for studies of how hearts function. Because the Pacific hag (*Epatretus stouti*) has no heart nerves or other sympathetic nerves, researchers have gained information they could not get from other animal groups.

Although of ancient lineage, lampreys and hagfishes are not typical in external appearance of most of the jawless fishes that lived about 400 million years ago in the Silurian period. (This period has been called the Early Age of Fishes.) Nor do they resemble the jawless fishes of the Devonian period that followed. The Devonian jawless fish was a heavily armored creature that lived on the bottom of lakes and rivers and was less than one foot long. They are appropriately called ostracoderms (Greek *ostrakon* = shell and *derma* = skin). Just why they were armor-plated is a mystery. Some experts think it was to protect them from giant scorpionlike creatures up to seven feet long. These ancient, mainly flat-bodied fishes were sluggish creatures. Lacking jaws to seize prey, they were grubbers in the mud, sucking in small animals and food particles. They disappeared toward the end of the Devonian, giving way to more efficient fishes that evolved during the period. These fishes had jaws with which to seize prey, and bodies shaped and powered for quick movement. They had paired fins to balance and control the swift dash at the prey.

"The evolution of jaws must have been the most far-reaching single event in the history of fishes," writes the British zoologist N. B. Marshall in *The Life of Fishes*. In the Devonian period—from about 390 million to 340 million years ago—it led to a veritable population explosion of fishes—so much so that the period is commonly called the Age of Fishes.

In the next chapter we'll look at the modern descendants of the first jawless fishes. These are the sharks, skates, rays and chimaeras (ratfishes or rabbitfishes).

6

SHARKS

WHEN I WAS a small boy we used to search the beaches after storms for what adults told us were mermaids' purses. They were empty capsules of a horny material and up to eight inches long; some were elongated and flat, but others were graceful cylinders with concentric spiral frills. They were, of course, the empty egg capsules of dogfishes (*Scyliorhinus* spp.) and of the Port Jackson sharks (*Heterodontus* spp.), some of the most ancient of all sharks—so ancient, indeed, that sharks of this genus are probably little different than they were 180 million years ago. Their eggs, also archaic, are among the most wonderful laid in water. They have a beautiful, concentric, double-spiral frill. The sharks that lay them are called Port Jackson sharks because they were first observed for science in the waters of Sydney Harbour. They are also found in parts of the Indian and Pacific oceans. The Californian species, *Heterodontus californicus*, lays single eggs during February and March.

Such large eggs are really not typical of sharks—or of

bony fishes. Most sharks bear live young, and most eggs laid by the bony fishes are small. (The bony fishes have skeletons of bone, whereas sharks have skeletons of cartilage. There are other differences between sharks and bony fishes, too, which we'll discuss further on.)

Incubation of the eggs of dogfishes and Port Jackson sharks is prolonged, running to many months—seven months in the case of the Californian example of the Port Jackson shark. The eight-inch-long youngster that emerges from the capsule is a miniature replica of its parents and able to fend for itself from the start. Incubation of the eggs of the European dogfish (*Scyliorhinus caniculus*) takes from 22 to 25 weeks. Some of the smaller dogfishes lay two to ten eggs at a time.

Some capsules laid by sharks have long tendrils, sometimes up to seven feet long, which cling to rocks and seaweed and anchor the eggs. Skate eggs are flat and oblong-shaped with a pair of pointed "horns" at each end. These horns are hollow so that seawater can enter to aerate the eggs. The undersides of the eggs are sticky and adhere to stones, sand and seaweed.

Most sharks that lay eggs are comparatively small, but the egg-layers include the largest of all living sharks, the whale shark (*Rhincodon typus*), which has been measured at 50 feet and probably reaches 60 feet. Its large, yolky eggs may be one foot long. The whale shark is a harmless creature, feeding on plankton in the warmer seas. Not much is known about its breeding habits, but 16 eggs were found in a captured one. Another large shark, the Greenland shark (*Somniosus microcephalus*), up to 24 feet long, lays eggs the size of goose eggs, in large numbers.

Some of the largest eggs, up to five inches long, are produced by the frilled or eel shark (*Chlamydoselachus anguineus*). These deepwater sharks of ancient lineage incubate their eggs, ten to fifteen at a time, inside their bodies, and carry them for up to two years before the pups are born. The frilled sharks have slender, eellike bodies up to seven feet long, and large mouths filled with sharp teeth. Fishermen, finding these sharks dead after storms, have sometimes claimed they've found "the sea serpent." It is just possible that very large specimens of frilled sharks have given rise to some of the more credible stories of sea serpents.

Altogether there are some 300 species of sharks, rays, skates and chimaeras making up the class of Chondrichthyes, or cartilaginous fishes. (The name is derived from the Greek *chondros* = cartilage, and *ichthys* = fish.) They are sometimes referred to as the Elasmobranchii or elasmobranchs.

Sharks and rays were once considered "primitive," on the assumption that a cartilaginous skeleton represented an earlier rung on the evolutionary ladder than a bony skeleton. Recent research suggests it is not as simple as that. The earliest vertebrates began with cartilaginous skeletons and then developed bony ones; some of their descendants, the sharks, took one path (cartilage), and other descendants, the bony fishes, another. Another major difference between the two kinds of fishes is that sharks, unlike bony fishes, have no swim bladder. Thus they cannot float motionless in the water but must keep swimming if they are to avoid sinking to the bottom. Still another difference is that, instead of a single gill cover, sharks have one for each gill. Thus, there are sharks with five, six or seven gill openings. Sharks lack the usual fish armor of bony plates or thin, overlapping scales; instead they have a covering of skin *denticles* (sharp, toothlike projections).

Almost all sharks live in salt water. A few species live in freshwater—in the Ganges and Zambezi rivers and in Lake Nicaragua. They inhabit the seas from the shallows to the depths—some to 10,000 feet.

Sharks and their relatives are all carnivorous, and, unlike most true fishes, all fertilize their eggs internally.

Male sharks and rays have a pair of cartilaginous appendages called "claspers" or intromittent organs which are introduced into the female's vent during mating. Each clasper is a kind of inner extension of the pelvic ("hip") fins and has a groove on the inner side to carry the sperm. When copulating, one clasper only is introduced. Few biologists have been lucky enough to witness mating among sharks and rays, thus little is known about the courtship of these creatures. With one pair of dogfishes that were observed, however, copulation lasted for 20 minutes.

Those sharks and rays that give birth to live young are called *ovoviviparous* and *viviparous*. In the first instance, the embryos develop inside the thin capsules by, first, feeding on the yolk and, later on, feeding on a special fluid secreted by the mother's uterus. The majority of sharks and rays are ovoviviparous. In the second instance, the embryo's yolk sac, after its nutriment has been absorbed, comes into contact with the mother's uterus, and her blood supply becomes linked to that of the embryo.

The numbers of live young born vary widely from species to species. As a general rule, the smaller sharks give birth to only a few at a time, while the big scavenging sharks, such as the tiger and hammerhead sharks, may bear at least 20 and often many more. (Eighty-two young were found in a captured tiger shark, *Galeocerdo cuvier.*) One of the myths of sharklore is that mother sharks swallow their young to protect them when danger threatens. It's not true, of course, but it's easy to see how the myth started. Fishermen have cut open captured, live-bearing sharks and found active young. (A similar tall story is told about snakes, and for the same reason—some snakes give birth to live young.)

What may be the most spectacular method of giving birth is that of the devil rays or devilfish (Mobulidae). Devil rays are so named because they have two fleshy horns on either side of their mouths. Devil rays have a habit of leaping high out of the water and falling back with a resounding smack. It has been generally believed that these acrobatic leaps are attempts to shed sea lice, but recently observers have noticed that newborn young are sometimes ejected during the leaps. It could be that the leaps are made to give birth to young.

There's a popular belief that all sharks are scavengers which will eat anything, including carrion. This isn't true; most sharks are selective, and the big ones prey mainly on live fish. The dark-barred tiger sharks, however, can be described as "living garbage cans." One caught in the Pacific had in its stomach bits of soap, cigarette packages, six empty bottles and pieces of metal. From the belly of another caught in Sydney Harbour, fishermen recovered three unopened bottles of beer. From another was recovered a small bag of coal and a set of false teeth.

Both sharks and more modern fishes have common ancestors in the first jawed fishes which developed toward the end of the Silurian period. Some of these placoderms, as they're called, were as fearsome as the most savage of today's sharks. There was, for instance, the aptly named *Dinichthys,* meaning terrible or huge fish. Some of the best fossil specimens of *Dinichthys* were excavated from the center of Cleveland, Ohio. *Dinichthys,* the scourge of the smaller placoderms and the evolving sharks and bony fishes, was 20 to 30 feet long. One incomplete fossil was estimated to have been 40 feet long and was named *Titanichthys. Dinichthys* had an armored head shield three feet long, and huge gape (of fully opened jaws) with jagged projections serving as teeth. The creature's jaws worked in a very odd way. Most animals open and close their mouths by lowering and raising the bottom jaw; *Dinichthys* had a head hinged to its body in such a way that it could both raise its head and drop its lower jaw. Thus it could take a very large bite. Not all placoderms were as fearsome as *Dinichthys;* most were small fishes from one to a few feet in length. Some fed on algae and invertebrates—and on each other.

You will have observed that sharks and rays that lay eggs or produce live young take no care of them afterward. No shark parent guards or feeds its young in the way some creatures do. The young have to take their chances. For the main part, as we saw with the lampreys, fishes find safety in numbers; they lay many eggs, sometimes running into millions, so that a sufficient number will reach maturity to carry on the race.

From their placoderm ancestors, the higher fishes took two widely diverging paths. The sharks and rays moved into the seas and, finding abundant food in the marine invertebrates, developed in the course of evolution into the forms we know today. The bony fishes (Osteichthyes),* as they're called, developed first in freshwater before they invaded the seas. The name Osteichthyes distinguishes the more modern fishes from the more ancient sharks and lampreys in which the ancestral bone skeleton has been lost. And a bony skeleton has been an essential factor in the evolution of vertebrates, both in the seas and on land. On the whole the higher bony fishes have been more successful than the sharks and rays. There are many more of them and many more species—about 20,000 known species of bony fishes as against 300 species of sharks and rays. The bony fishes are masters of the water and have been the dominant fishes for about 180 million years. Nonetheless, we shouldn't discount sharks and rays too much. There were sharks in late Devonian times and there are sharks today. All the major families of today's sharks and rays had developed as long as 60 million years ago. They have adapted to many changes over that immense period of time and, although they no longer dominate the seas as they did in past ages when some sharks grew to 50 feet in length, they look like they'll continue to inhabit the seas for a long time to come.

---

* From *osteon* = bone and *ichthys* = fish.

7

LUNGFISHES AND COELA-
CANTHS: THE LIVING FOSSILS

SHOULD YOU come upon the fish which is the subject
of this chapter in an aquarium, it is probable that
you will not be impressed. Someone once described it as "an
uncommonly dull, lazy fish." You can touch it and get only a
slight response; it may swim a foot or so away and then remain
stationary like some giant tadpole. This fish is four or five feet
long—some grow up to six feet—and with its dull green sides
and back and its long flat tail, it looks not unlike a rather
sluggish conger eel. The scales are large, and its two pairs of
fins are worth a second glance; they are not ray-shaped like
those that we have come to know in most fish, but instead are
fleshy lobes, rather like short paddles. In the aquarium it
probably won't perform for you its trick of rising to the surface
and gulping in air. In its tropical home in southern Queensland, Australia, it does this every half hour or so, but in the
aquarium's well-oxygenated water it may have no need to
augment the oxygen which it absorbs with its gills from the
water.

The fish is, of course, the famous Australian lungfish, *Neoceratodus forsteri*, which we may roughly translate as "Forster's new horny-tooth." It is a "living fossil"—a term coined by Charles Darwin—and is almost indistinguishable from the fossil *Ceratodus* which was widespread over the world 300 million years ago. You can imagine, therefore, the excitement of the then curator of the Australian Museum in Sydney, Mr. Gerhard Krefft, when just over 100 years ago, in January 1870, he unpacked some salted specimens of Australian lungfishes and examined them closely. "I never saw anything to equal this in my life," he said afterward. There before his eyes was evidence that a fish believed to have become extinct 100 million years ago was still in existence.

Behind the naming of this living fossil lies a fascinating story. *Neoceratodus* means, of course, new *Ceratodus,* and *forsteri* honors the man who made the discovery possible. This was a Mr. William Forster, a New South Wales cabinet minister who had earlier been a rancher in southern Queensland. He was a friend of Krefft's and told him about the strange and marvelous "fresh-water salmon" with one lung in the waters of the Burnett River. Krefft could not believe that such a phenomenon existed. But so persistently did the story of the unusual Queensland fish crop up during the friendly visits made by Krefft to the Forster home in Sydney that at last, in 1869, Krefft challenged William Forster to produce some of the marvels. Accordingly, Forster wrote to his nephew, a Mr. M'Cord who lived in the Burnett River area, and asked him to catch, salt and send to Sydney some of the "Burnett salmon," as they were called there because of their red-colored flesh.

Krefft's startled announcement in a letter to a Sydney paper became world news. Museums at first paid ten pounds —worth about U.S. $50 and a considerable sum in those days—for specimens, but as the fishes were plentiful in the Mary and Burnett rivers, the price fell rapidly to ten shillings.

Live specimens, too, were sent abroad to England and the United States. Lungfish are hardy and long-lived—they live to, possibly, 100 years. One in the Zoological Gardens in London lived for nearly 20 years, and another in America lived for 30. (Knowing that his lungfish pet is likely to be long-lived, a Sydney fish fancier has provided for its maintenance in his will.)

If we were to keep a male and female Australian lungfish in an aquarium and they were to breed, we might not think the eggs very remarkable. They are externally fertilized, in the way that most fish eggs are, and have a thick, gelatinous covering. They resemble those of frogs and are about one-quarter of an inch in diameter. They are not buoyant and sink to the bottom.

But if the eggs were to hatch, then we would note something remarkable enough to start us thinking. So similar is the embryo of the Australian lungfish to that of a developing frog (an amphibian) that it is barely distinguishable from it until a comparatively late stage when the characteristic large tail "fin" grows. And in the course of trying to raise the young Australian lungfish, we would discover something peculiar. We would find that we'd have little success unless we kept them in shallow water, about half an inch deep, and provided them with a small sandbank on which the fish could rest for a while *out of the water.*

Don't let the name "lungfish" mislead you into thinking that this extraordinary Australian fish can live and breathe by means of its lung alone. When removed from the water, it will die just as quickly as any other fish—just as soon as its four pairs of gills have dried out and ceased to function. These facts should permanently dispose of the myth—still widely believed in Australia and elsewhere—that the Australian lungfish can leave the water and crawl by means of its paddlelike fins. In front of me as I write is an article by an Australian naturalist who alleges that the Australian lungfish can exist on land as well as in water and "can slither along dry river beds, over sand, or even cross rocks in search of a nice new, watery home." One early Australian explorer, Carl Lumholtz, wrongly depicted the lungfish as basking in the sun on a log, much like a lizard.

What the Australian lungfish does have is a supplementary lung that does enable it to survive times of drought when other fish die by the hundreds and thousands. "You can see lungfish lying on the weeds and plants in the shallows, in just a few inches of water," said a Burnett River resident recently. "They can stay there for days."

Something else that might arouse our interest is that this sluggish Australian lungfish "walks" on its fins at times along the bed of its aquarium. And if we could dissect the fish, we would discover still other things to keep us thinking. For instance, the nostrils of ordinary fish are blind pits, but those of this fish lead into the mouth cavity and have much the same arrangement as four-legged, air-breathing animals. Looking at the single lung, we would discover that for all practical purposes it is an excellent lung whose blood vessels are arranged in much the same pattern as those of frogs and newts, which, of course, are amphibians.

A lung may seem a remarkable thing for a fish to have, but some other living fishes of ancient lineage have them, too, while other fishes have the remains of a lung or lungs. (We shall see why later.) This is, of course, the swim bladder, which acts much like the ballast tanks of a submarine and allows a fish to rise and sink in the water.

It is, perhaps, little wonder that the first man to describe the Australian lungfish, Mr. Gerhard Krefft, called it an amphibian. It is not, of course, although it could be a fish somewhere on the way to becoming an amphibian.

Herein lies the great interest of the Australian lungfish

and its distant cousins in Africa and South America. They are the descendants, somewhat specialized, of a vast group of fishes which lived over 300 million years ago. The first amphibians almost certainly evolved from some fishes in this group.

The amphibians we know today include the frogs, toads, newts and salamanders. (The name is a compound of two Greek words, *amphi* = both, and *bios* = life, and is given to these creatures because they spend their lives in freshwater and on land.)

Let us imagine that we can board a time machine to take us back to the end of the Devonian period when the stage is set for the emergence of our first amphibians.

The Devonian period, which had existed for about 50 million years before we arrived, was a time of warm days when widespread droughts succeeded annual floods. (Some fossils of early amphibians have been found in ancient swamps in what is Greenland today.) When we disembark from our time machine, we set foot in a world in which there has been a great expansion of life—particularly of new fishes and land plants.

The land plants strike us as strange. None has flowers, and some are leafless. Most striking of all are the great naked forests of scale trees, so-called because their trunks are covered with overlapping, leaflike "scales." Some of these trees with their scanty crowns are three feet thick and grow to 40 feet or higher. More familiar are the tree ferns.

Walking among the trees and plants, we disturb primitive spiders, millipedes, mites, tiny scorpions and other early insects.

We walk along the shores of warm, shallow seas filled with fish and studded with coral reefs. Life had originated in the seas, but the first great radiation of fishes was in freshwater in the Devonian period. From the rivers and lakes, new forms of fishes began to invade the sea. Not only was the Devonian period the Age of Fishes; it was also the Dawn of the Amphibians.

In addition to the jawless fishes (Agnatha), we discover many kinds of true or bony fishes (Osteichthyes). And we can watch the first sharks (Chondrichthyes) hunt their prey in the shallow seas. But the most fearsome predators are the placoderms, such as the 30-foot-long armor-plated *Dinichthys* with a gape of four feet.

With the hindsight of time travelers, we know where to look for the ancestors of the amphibians—in one of the two subclasses or branches of the Osteichthyes. From the Middle Devonian, when the first bony fishes emerged, there were two more or less distinct types of fishes. The first subclass, which does not concern us here but will later on in Chapter 8, was the Actinopterygii. These are the ray-finned fishes in which the fin developed fanlike rays anchored to a large base. Today, their descendants dominate the waters of the earth.

The other subclass or branch which interests us now is that of the Sarcopterygii (from *sarx* = flesh, and *pterygion* = little wing)—the fleshy-finned fishes.

When we look closely at some of these freshwater fishes with fleshy fins, we notice that some look vaguely familiar. For instance, those fish that rise to the surface of that stagnant pool to gulp in air are lungfish, ancestors of the lungfishes that persist today in Australia, Africa and South America. Biologists know them as the order Dipnoi (*di* = double, and *pnein* = to breathe). Are these fishes also the ancestors of the amphibians? In the past, experts used to think they were, and that the present-day survivors were in the direct line from fish to amphibian. But today's lungfishes have certain specializations which show that they are off the direct line of descent. They are first cousins, so to speak. They eat small mollusks, and their jaws are modified for this diet. They have specialized teeth fused into fan-shaped crushing plates. As Dr. A. S. Romer, the famous American paleontologist, puts it in *Man and the Vertebrates*, "The lungfishes are, so to speak, not the ancestors but the uncles of land-dwellers." And our own great-great-great-uncles.

Continuing our time traveling, let us look at another group of freshwater lobe-finned fishes, the order Crossopterygii (*krossoi* = tassels, and *pterygion* = little wing or fin). Those large fish with large, round, bright-blue scales and the two large lobe-shaped fins—could they be coelacanths? Well, the grandfathers of the coelacanths. When in 1938 the coelacanths were dramatically rediscovered in South African waters, the thing that astonished the experts was how little these fish had changed in an immense period of 300 million years.

But again with hindsight we know that the Devonian coelacanths (*koilos* = hollow, and *akantha* = spine) aren't the ancestors of the amphibians which are to emerge toward the end of the Devonian period. We must look for them in their close cousins the Rhipidistia (*rhipis* = fan). These are air breathers like the lungfishes.

Probably air breathing was practiced by all the major Devonian fishes except the sharks and chimaeras. They had to acquire this method of supplementing their oxygen if they were to survive. Because the climate of the Devonian period was either very rainy or very dry, just as it is today in certain tropical parts of the world, in the dry seasons the streams and ponds became a series of stagnant water holes. They daily grew fouler, losing more and more oxygen as time passed and the water evaporated, and the decaying dead bodies of the weaker fishes contributed their share to the general deterioration. Significantly, the three genera of living lungfishes persist in tropical regions where a short wet season is followed by a long dry one. What the Devonian bony fishes developed appears to have been a double sac, opening out from the underside of the chest, which allowed them to make use of atmospheric oxygen.

When the pools dried up, some freshwater fishes were

probably able to survive in cocoons of mud the way South American and African lungfishes do today. Fossilized remains of burrows made by lungfishes have been discovered in Paleozoic sediments in Texas. Some burrows even contain the remains of lungfishes that failed to escape from their mud prisons.

Other fishes found another way of surviving. For example, it is reasonable to suppose that under the trying conditions of severe and prolonged droughts some Devonian rhipidistians were able to waddle, with the aid of their stout, well-muscled fins, from an almost dried-up pool to another with more water. The fossil record tells us how the first primitive amphibians must have evolved from these "walking" fish.

We can speculate that those rhipidistians which were able to stay longer out of the water would have survived when the others died. And they would have had progeny with their own greater ability to breathe air. They and their descendants would have found plenty of food on land—stranded fish, insects, land plants.

Importantly, the rhipidistians had the physical equipment that could evolve into limbs. They had within the fleshy lobes of their two pairs of fins the bony skeletal supports which could evolve into limbs. One bone in the lobe was attached to the shoulder skeleton; linked to this bone were two others which formed a second fin joint. From this joint a series of bones branched out, much as our fingers do. Here were the beginnings of the pentadactyl (five-fingered) limb of reptiles, birds and mammals. In the course of evolution, the lobe fins developed into serviceable legs for amphibians so that they could walk on land, or even climb, or handle objects.

This method of survival by waddling from one drying-up pool to another wasn't available, of course, to the other major group, the actinopterygians, with their frail, fanlike fins.

Probably the ancestor of the amphibians was a rhipidistian called *Eusthenopteron** which closely resembles the first amphibians in the bones of the skull, in the skeleton of its fins and also in the teeth, whose enamel covering is folded in toward their centers. *Eusthenopteron* was up to two feet long and preyed on other fishes. Superficially, *Eusthenopteron* looked not unlike a coelacanth.

The fossilized remains of one of the first amphibians have been found in Greenland. *Ichthyostega* has a fish tail, lungs and well-developed legs and feet. All the primitive amphibians were like that—with fish tails and short, stubby limbs that could barely lift their bodies off the ground. They must have spent much of their time in the water. The paradox is that the first amphibians didn't develop lungs and limbs so that they could leave the water but in order to continue to live in it during seasonal droughts. "Land lungs were developed to reach the water, not to leave it," says Dr. Romer.

From these beginnings a way of life on land slowly developed. The rhipidistians were predators, but one of the ironies of evolution is that the rhipidistians—which had become extinct by the end of the Permian period—were probably themselves eliminated by their own predatory amphibian descendants.

The move to the land was one of the most difficult ever undertaken by the vertebrates. As fishes the weight of their bodies had been partly supported by water. On land they had to bear the full weight. The emerging tetrapods (four-legged creatures) had to develop stouter spines and strong limb muscles to support their bodies. The first amphibians were badly bowlegged; some could scarcely lift their bodies off the ground. In addition, they had to acquire eardrums to pick up sounds (air is not as good a conductor of sound as water is), and new kinds of eyes to see on land.

The environment pressed ruthlessly on all the fishes in the

---

* Literally, "stout fin."

Devonian period and in those that followed. Some took to the land, as we have seen; others became extinct when they failed to adapt quickly enough to changing conditions. Some, such as the lobe-finned fishes and the lungfishes, began the long journey to near-extinction.

Of the living lungfishes, the Australian species is considered the more primitive. Certainly its African and South American cousins have made better adaptations to living in substandard conditions. The African lungfish, which rather resembles the newt in general appearance with slim fins, must be one of the hardiest fishes there are. It can live in little water of the foulest character, getting about 95 percent of its oxygen from the air by means of its two lungs. It must because its gills are vestigial and it will die if denied access to the air—as an American zoologist, Dr. Homer W. Smith, discovered. He tells in his book, *Kamongo*, how for a long time his efforts to keep adult African lungfishes alive failed because he kept them in too-shallow tubs of water. They couldn't tilt their bodies enough to get their nostrils above the water to take in air, and drowned.

Dr. Smith, who is one of the world's experts on the African lungfish (*Protopterus* spp.), relates in *From Fish to Philosopher* how he moved 150 fishes to New York from Lake Victoria. The fishes, each about one foot long, were carried in five-gallon cans, some of which contained mud with almost no water and others about six inches of water with some mud. Two or three fish were placed in each of the cans. The trip took six weeks, and the fish were stored in these cans for about twelve weeks altogether. The distance traveled was over 8000 miles. The facts of the trip, as Dr. Smith said, all testify to the remarkable hardihood of the African lungfish. They suffered repeated splashing and jarring. Those in water endured many changes of water, including slightly chlorinated water at Nairobi, ship's distilled water at sea and miscellaneous water taken at various ports. They survived very high and prolonged temperatures in the Red Sea. One can of mud with three fish fell several feet and was turned upside down in a French baggage car. The fish were repeatedly poked at by customs officials and others. During the whole time they were not fed, because Dr. Smith thought that they would travel better if putrefactive contamination of the mud and water was kept to a minimum. "In short," he says, "they suffered an ordeal which few other animals and certainly few fishes could have endured." And they survived.

Incidentally, African lungfishes, of which there are four species, grow up to seven feet long, and weigh about 100 pounds.

Not only can the lungfish live in little water, the African species and the South American species (*Lepidosiren paradoxus*) are able to survive for months or even years when the rivers that they normally inhabit are completely dried up. When the dry season starts, these fishes burrow into the mud. The fish's wriggling shapes a bulbous cavity in the drying mud. When the water finally evaporates, the fish is left curled up inside a mud cell with only a small blowhole opening to the air. Its tail is across the top of its head to protect its eyes and to cut down on evaporation by reducing the exposed surface. The fish's body becomes covered with a slimy mucous secreted by the skin. When this mucous dries, a brown, parchmentlike, waterproof cocoon completely envelops the body except for the breathing opening. Safe in its cocoon, the African and South American lungfish can live for up to four years in this hibernationlike state. (The term used by biologists for this is *aestivation* and not hibernation. "Aestivation" derives from the Latin *aestivare*, which means "to pass the summer.") Dr. Smith found that the metabolic rate of an aestivating fish was so slowed that the intervals between breaths extend to one or even several hours. (An active African lungfish breathes at the surface of the water at least every 15 minutes.) Dr. Smith also found that the heart slows to only about three beats a minute, and that except for the infrequent pulse and respiration, the fish appeared to be in a state of suspended animation.

In some experiments African lungfishes were persuaded to aestivate in mud and in plaster of paris. The plaster was just as effective as the fishes' native mud, but, of course, the biologist had to use a chisel to free them. The fishes survived at least two years in mud or plaster, and one survived three years of aestivation and a further year without food when returned to the water.

While the African lungfish is aestivating it lives off its muscle tissue. If the period of aestivation is prolonged to two or three years, then it is a lean and desiccated-looking fish that emerges finally—looking much like a piece of old leather.

Dr. Smith described a freshly liberated African lungfish as struggling with grotesque movements to the surface of the water in order to breathe. Several days passed before the liberated fish absorbed enough food and water to regain its former appearance. Dr. Smith said that for several days the stiffened and twisted body looked very much like an animated horseshoe trying to stick one end of itself out of the water and that for days the swimming movements were erratic and uncoordinated.

What astonished Dr. Smith was that the African lungfish bred shortly after it emerged from aestivation, so that it had, in fact, scant time for recuperation. But obviously the ability to breed without undue waste of time is of survival value to a fish living in areas where rain is infrequent. The quicker it breeds the greater the chance of descendants. And it had survival value in the remote Devonian past of this ancient fish.

In breeding, the female makes a crude nest in weeds or swamp grass in shallow water. There she deposits her eggs, which are externally fertilized by the male. The male stands guard while the eggs are incubating, which takes about eight days, and while the larvae remain in the nest. All the time he continually swishes the water with his tail to improve aeration. A young lungfish has a sucker organ, rather like that of a tadpole, with which it attaches itself to the side of the nest. This device probably assists them greatly because they can rest while absorbing their heavy yolk. Moreover, by keeping still, they are less likely to be seen by enemies. The larvae develop feathery external gills which they do not lose until they are approaching maturity. The large surfaces of these temporary gills allow the larvae to make the most of the limited amounts of oxygen in swampy water. The larvae of the South American lungfish also have temporary gills.

The breeding of the South American lungfish greatly resembles that of its African cousins. The fish spawn in deep holes, and again the nest is guarded by the male, whose pelvic fins grow temporary gills. Like his offspring with their temporary external gills, the male parent is thus enabled to capture the limited amount of oxygen in swampy waters; he need not leave the precious eggs or larvae and swim to the surface to gulp air, as he normally does.

As we have seen, all the modern survivors of the ancient lungfishes in Australia, Africa and South America made remarkable adaptations to difficult conditions. In stagnant water they were able to breathe, and some acquired the ability to lie dormant for months or years when the rivers and streams dried up. This was all very well, but, as Dr. Smith points out, it was a better way of life to stay awake all the year round by crawling from one water hole to another. This was the solution that slowly evolved with the development of the first amphibians.

Until recently this dramatic development could only be studied in fossils. Then, in 1939, fishermen off East London, South Africa, hauled up the first of a number of coelacanths—

the sole known survivor of the ancient crossopterygians. As we know, the coelacanths did not give rise to the amphibians, which are a bit further up the evolutionary ladder. But, nonetheless, biologists can now study the lobed fins, particularly the soft parts, and probably gain some clues how these fins developed into amphibian limbs.

The discovery of this second living fossil is one of the great stories in natural history of this century. The men on the trawler off the mouth of the Chalumna River in South Africa in December 1939 knew they'd found something unusual. Although they did not know it, their experience was as unique as if they'd walked down the street and been confronted with a dinosaur. The fish was five feet long, weighed 127 pounds, had large, round bright-blue scales and dark-blue eyes. Importantly, it had two large, lobe-shaped fins. (See Figure 3.)

The fishermen reported the strange find to Miss M. Courtenay-Latimer, Curator of the East London Museum. Miss Courtenay-Latimer was unable to identify the specimen and sent a sketch and descriptions to Professor J. L. B. Smith of Rhodes University College, Grahamstown, South Africa. By ill chance, her letter was not delivered for ten days, during which time the body of the fish rotted and had to be destroyed. Only the head and the skin were saved by a taxidermist in East London.

Professor Smith, suitably startled by the sketch and description, tentatively identified the fish as a coelacanth, a survivor of the crossopterygians. He confirmed this identification when he saw the head and skin. His excitement must have been as great as that of Gerhard Krefft in Australia when earlier he had been confronted with the dramatic survivors of the lungfishes, thought to have been extinct for 100 million years. He named the fish *Latimeria chalumnae* to honor Miss Courtenay-Latimer and the place where it was found, then he started a sustained search for further specimens. After 14 years

FIG. 3.   *Coelacanth—a "living fossil."*

to the very month, a second coelacanth (named by Professor Smith *Malania anjouanae*) was caught off Anjouan in the Comoro Islands, 200 miles west of Madagascar. Since that time a number of living specimens, from 70 to 180 pounds, have been caught in waters near the Comoro Islands. The fish have been captured at depths from 480 to 1200 feet, but Professor Jacques Millot, director of the Institute of Scientific Research at Madagascar, thinks that the fish lives at a depth of 2400 feet.

What fascinates biologists about *Latimeria* is its slow rate of evolution. It differs from its Devonian ancestors and later forms only in minor details.

Over a long history of 300 million years the coelacanths spread over most of the world, for the main part in freshwater ponds, lakes and streams. A few of them took to the sea, but most of them continued in freshwater until the Triassic period.

No one can say how long *Latimeria* has been a marine fish. Nor is anything known about its spawning habits; we don't know whether it lays eggs or whether it is ovoviviparous. But we have at least a clue in a single coelacanth fossil from the Jurassic period. It is a full-grown fish containing the skeletons of two small coelacanths. One biologist who has studied this fossil closely thinks that the small fishes were not eaten by the full-grown fish but are unborn embryos. If he is correct, then this Jurassic coelacanth was ovoviviparous.

As you already may have guessed, this fish of the depths, where there is little or no sunlight, no longer breathes air.

These "living fossils" from the Age of Fishes—the lungfishes and the coelacanths—so little changed over 300 million years, may yield some clues on how fishes left the water and began the conquest of the land. Therein lies their extraordinary interest, even if one biologist found an Australian lungfish "an uncommonly dull, lazy fish."

A pleasing footnote to the dramatic discovery of the Australian lungfish is that the Queensland government has honored this ancient fish by giving the generic name of *Ceratodus* to a railway station. And a township on the homeland river of the fishes is called Theebine, one of the Aboriginal names for the *Neoceratodus*.

8

FIRST MODERN FISHES

WHEN I WAS about to write this chapter, I attended a party where people were eating caviar canapés, and the odd thought crossed my mind that we were eating some of the most ancient fish eggs it's possible to eat. By that I mean that sturgeons have not changed much from their primitive ray-finned ancestors, creatures that appeared in the middle of the Devonian period. The red and black roe we were eating had come from seagoing sturgeons such as the Caspian and Black Sea beluga (*Huso huso*) and other related species from Russian waters. Both these and the Atlantic sturgeon (*Acipenser sturio*) grow very large, up to 25 feet. The more general range of size for the 30-odd species that inhabit seas, large rivers and lakes in the Northern Hemisphere is from 5 to 12 feet long. Typical sturgeons spawn in freshwater, much like salmon, and then go to sea to feed and reach sexual maturity before returning to their home waters to breed. But some species are confined to freshwater.

Most sturgeons show their ancient ancestry in various

ways. They have broad, armor-plated heads. They have one to several pairs of barbels. They have large, tilelike, bony scales along their backs and sides. Their fins have remained more primitive than other bony ray-finned fishes of ancient lineage, and their tails strikingly resemble the heterocercal (asymmetrical) tails of sharks, which are also of ancient lineage. The tip of the body tilts up into the upper part of the two-lobed tail—a primitive form of tail on the evolutionary ladder. A major difference from their primitive ray-finned ancestors is that they possess a typical air bladder rather than a lung.

In the course of evolution, sturgeons have degenerated; they have lost many of the scales they once had, have small mouths with weak jaws and are, for the most part, grubbers in the mud for food. Their menu consists mainly of small invertebrates, though some of the larger specimens feed on fish.

Close relatives of the sturgeons are curious paddlefishes of the Mississippi River and China. The Mississippi paddlefish (*Polyodon spathula*) is a slow-moving and slow-growing fish that reaches a weight of up to about 150 pounds and a length of six feet or more. Its Chinese cousin *Psephurus gladius* grows to about 11 feet and is found in the Yangtze River. Both these descendants of fish that emerged in Middle Devonian times feed on minute animals. They're easily distinguished from sturgeons by their remarkable long, spoon-shaped noses and their smooth, unarmored skins. The American species is commonly called the spoonbill—for obvious reasons—and it puts this long nose to good use. It is richly endowed with nerves and sensory organs to locate its food of water insects and shrimp. The fish swings its snout from side to side, stirring up the mud at the bottom of the river and disturbing its prey. The Mississippi paddlefish opens its large mouth and engulfs food and mud together. Fortunately, it is equipped with efficient gill rakers which act as sieves. They hold back the water fleas, etc., and let the water and mud pass through.

These fishes are sexually mature when they are about four feet long. They are caught by commercial fishermen using nets and lines, and their eggs are sold as caviar.

The ancient ancestry of the sturgeons and their cousins, the paddlefishes, is underlined by the name biologists have given to the superorder to which they belong—Chondrostei (*chondros* = cartilage and *osteon* = bone). The key word is, of course, "cartilaginous," recalling even more ancient fishes, such as the placoderms, which were the ancestors of both the bony fishes and the sharks. Sturgeons have mainly cartilaginous skeletons but bony jawbones. Importantly, they have notochords, those axial supporting rods that are replaced by a vertebral column in most advanced fishes. The notochords are surrounded by partly segmented cartilaginous vertebrae and, incidentally, are prized as a sagolike (starchy) food in the Soviet Union.

The sturgeons and paddlefishes thus have more to commend them—to paleontologists, anyway—than the caviar delicacy they produce. They and some tropical African fishes called "bichirs" are relics of the first ray-finned fishes. It is thought they developed from freshwater placoderms, whereas sharks are descended from saltwater placoderms. The sturgeons take first place, even if they are relics, as the largest of all freshwater fishes. (It could be argued that, strictly, the largest freshwater fish is the giant arapaima, *Arapaima gigas,* of South America, which is said to reach 15 feet. More conservative estimates are up to 8 feet.) Sturgeons have survived in freshwater, for the most part, because competition with more advanced fishes is less savage there than it is in the sea. You may point out that most sturgeons live in the sea, but what is more important is that they breed in freshwater. And reproducing themselves is, of course, the essential matter.

The common or Atlantic sturgeon enters rivers on both

sides of the Atlantic in spring in order to spawn. Females lay their eggs in early summer, forcing the eggs out by rubbing their bellies on rocks. The gray-colored or black eggs are about a ninth of an inch in diameter and are adhesive. Each female lays from 1.5 million to 3 million eggs. For those with an interest in caviar, a female weighing about 1000 pounds will lay eggs the total weight of which will be slightly over one-fifth of her own weight—or about 224 pounds. The beluga sturgeon swims from the Caspian Sea to spawn in deep holes in the Volga River.

The newly hatched larvae of sturgeons are about half an inch long. A striking feature is a shallow, pigmented groove just in front of the mouth—a relic of the sucking disc of its primitive ancestors and one that it shares with the larvae of other primitive fishes such as the bowfin (*Amia calva*), gars and lungfishes.

In Russia the females are captured, stripped of their eggs and then released. At some Russian stations in the river mouths where the females, heavy with roe, are herded into enclosures in the shallows, up to 20,000 sturgeon are sometimes caught in a single day. After stripping, the eggs are salted to preserve them.

In recent years the sturgeon fisheries have been overexploited, and the amount of roe exports has declined. But the Soviet Union is now taking active steps to prevent the loss of this luxury food. The common sturgeon was once common in the Hudson River. But pollution and overfishing have reduced its numbers from former days when its tasty red flesh was frequently on sale and was jokingly called "Albany beef." A record 18-foot sturgeon was once caught in New England waters, and there was also extensive fishing on the West Coast of the United States. In 1872 up to 5000 sturgeon a month were caught in the Sacramento River, California. Inevitably exploitation led to a decline and, ultimately, to protection in

1917. Since then the sturgeons have been coming back. In British waters the common sturgeon is still a royal fish—that is, the Queen has first right to any caught.

After spawning, common sturgeons stay in the rivers and lakes until September when they return to the sea. Their *fry* (young) stay behind in freshwater. When they are a year old they, too, go to sea.

All sturgeon species are long-lived and all take a long time to reach sexual maturity—as much as 20 years in some instances.

One of the best-known North American freshwater species is the rock or lake sturgeon (*A. fulvescens*). Once plentiful, its numbers have been much reduced. It spawns in shallow water and rarely descends to the sea. It grows to about six feet.

We could eat the roes of some even more primitive living ray-finned fishes than the sturgeons if they weren't so difficult to gather. Little is known about the spawning habits of those tropical African fishes called "bichirs" (*Polypterus* spp.) except that they move into swampy water with the onset of the rainy season. Some ten species of these slender fish live in swamps and reclaimed margins of rivers and lakes.

The African bichirs closely resemble the oldest fossils of ray-finned fishes, and thus give us an even better idea than sturgeons do. They have thick and shiny bony scales and a series of finlets on their backs supported by a single spine, and they have paddle-shaped pectoral fins. The only living relative of the bichirs is the snakelike reedfish (*Erpetoichthys calabaricus*), a West African freshwater fish.

What makes both the bichirs and reedfishes unique is that alone among living ray-finned fishes they have well-developed, functional lungs opening from the underside of the throat. There is good reason why they sould retain their lungs, since they inhabit the upper Nile and tropical African regions where seasonal droughts occur, conditions similar to those of Devonian times.

Some changes from the ancestral type have occurred in the fins, such as the series of sail-like finlets along the back which give the fishes their generic name *Polypterus* (*poly* = many, and *pteron* = wing). An even more striking change is that the pectoral fins have developed into fleshy lobes. (Because of these lobes, the experts for years included *Polypterus* among the crossopterygians.)

The acquisition of fins with a fleshy lobe makes one wonder if these ancient African fishes are about to try to become amphibians. No doubt in the many millions of years of their existence in stagnant water there has been a strong incentive for them to take to the land in order to get from one pool to another. Ray-finned fishes, as we observed earlier, existed in Middle Devonian times, but with such fins they had little chance of successfully venturing onto the land. Their fins were first class for swimming, but far too feeble to support their weight out of water. Few ray-fins can crawl even a few inches. Some ray-finned fishes today do venture onto the land—these include the mudskipper (*Periophthalmus* spp.) and the Indian climbing perch (*Anabas testudineus*)—but, in both, the fins have changed into not very successful crawling or climbing tools.

You will recall from Chapter 7 that the Devonian bony fishes took either one of two important directions. In one group the primitive fin of the placoderms and acanthodians developed into a paddlelike or lobelike structure. These are the fishes that are lumped together in the subclass Sarcopterygii, and include the orders Crossopterygii and Dipnoi. In the Devonian period they were for long the dominant fishes. Some made the progress to the land and developed into amphibians.

The other subclass developed fanlike or raylike fins and appeared in the Middle Devonian. Called the Actinopterygii (*aktinos* = ray, *pteron* = wing, *pterygion* = little wing or fin), they are the common fishes of today, while the other group are extinct or almost so. Among the actinopterygians were the chondrosteans, the ancestors of the bichirs, sturgeons and paddlefishes. They multiplied during the Triassic age and dominated the waters, but in the Jurassic age that followed, more advanced ray-finned fishes developed and took over. A few chondrosteans hung on into the Cretaceous period and survive today.

The first of the more advanced ray-finned fishes were the holosteans (Greek *holos* = whole, and *osteon* = bone), which had more completely ossified skeletons, shorter jaws, more symmetrical tails and thinner scales than their ancestors. The gars and the bowfin (*Amia calua*) are the sole survivors today. The holosteans in turn gave way to the teleosts (Greek *telos* = end, and *steon* = bone), which were, so to speak, an improvement on the earlier models.

The advanced ray-finned teleosts have invaded almost every watery habitat. In the evolutionary race, the lobe-finned fishes and the sharks fell behind. Sharks were mainly carnivorous predators or scavengers and are so today. Bony fishes, as well as becoming predators and scavengers like the sharks, developed many omnivorous types.

The teleosts developed a protean variety of forms to fill almost every niche. Some, such as pike and trout, swoop on their prey; others, such as the sluggish anglers, lie in wait for it; others, such as the herrings, feed on zooplankton, and others, again, such as the milk fish, live on plants.

It's difficult to single out any single factor for the success of the teleosts. They had a number of advantages. They had more efficient fins and tails and were better swimmers than their lobe-finned cousins and the holosteans, and the sharks. Typical upturned, sharklike tails weren't as efficient as the homocercal (symmetrical) tails of the teleosts or higher bony fishes. The asymmetrical tail of fossil and living sharks causes a lift of the hind part of the body, which has to be countered by the paired fins; in the teleosts the paired fins are freed from this function. The backbone of a shark extends to the tip of the tail; in bony fish it ends where the tail starts.

What may have also helped them was a switch from the sense of smell to that of sight. Most lobe-finned fishes were mainly dependent on the sense of smell for their knowledge of the water world in which their survival rested. In the teleosts the organs of smell are comparatively unimportant compared to those of sight. "The eyes tend to be large and apparently are the dominant sense organs," says Dr. Romer, adding that this contrast between them and their lobe-finned cousins is shown in the brain organization. The transformation of their air sac or lung into a swim bladder may have helped the teleosts. In an advanced ray-finned fish the swim bladder which contains gas is a hydrostatic organ—that is, it takes in and expels gas and allows a fish to vary the specific gravity to match the depth at which it is swimming and thus float motionless in water. Advanced fishes are in effect almost weightless. Sharks, on the other hand, do not have swim bladders and can only rest on the bottom of the sea.

Then, too, though this is arguable, the teleosts could have peopled the seas by sheer weight of numbers. They "decided" presumably that reproductive safety lay in numbers. And swifter-swimming, ray-finned teleosts could colonize the seas faster than their slower cousins, the lobe-finned fishes and the lungfishes. "Quantity, not quality, has been the actinopterygian motto," says Dr. Romer. A codfish may lay up to 9 million eggs, but only two at most need to reach sexual maturity to replace her. The teleosts must be regarded as extremely successful. Their diversity is extraordinary—we know of more than 20,000 living species, and this number is added to yearly—and they have adapted themselves to almost every kind of water habitat, from tropical reefs to cold polar waters to the almost barren waters of high mountain streams.

The Osteichthyes make up the dominant class of our times. There are four other great classes of fishes, but, of these, two are extinct—the Placodermi (armored fishes) and Acanthodii (spiny fishes). The Agnatha (jawless fishes), after flourishing in Silurian and Devonian times, persist only in the lampreys, and the last class, the Chondrichthyes (sharks, dogfish, rays, skates and chimaeras), are in retreat. And within the Osteichthyes, the subclass Actinopterygii has outstripped the other subclass Sarcopterygii (lobe-finned fishes and lungfishes).

As a biologist once pointed out, 99 percent of all the species that have ever lived are extinct. The rest of Part One—and we're just over halfway—will be about the bony fishes and their eggs.

9

FISHES AND THEIR EGGS

REPRODUCTION AMONG most of the bony fishes resembles that of their more primitive fellows, the
lampreys. After a courtship, the female sheds her eggs, and the
male ejects sperm over them. In most instances, the eggs and
the emerging larvae are left to take their chance. The only
exceptions to this general rule are the live-bearers, the nest
builders and those that brood their young in pouches or in
their mouths or on their bodies. These are among the most
fascinating fishes there are and will be discussed later, in
Chapter 12. Males and females have gonads (sperm and
egg-producing organs) in long, paired sacs. The eggs and
sperm are generally discharged through ducts. You've probably seen them. The hard roes you see in fish shops are female
ovaries; the soft roes are male gonads.

Unlike the higher vertebrates, sex is not a simple matter
among fishes. Some are not invariably either male or female.
A few start life as a male or female and change sex as they
grow older. Some other fishes are hermaphrodites—that is,
each individual produces both sperm and eggs.

78

Among the fishes that change their sex are some sea perches belonging to the grouper subfamily Epinephelinae. As they grow bigger the males turn into females. The same thing may happen to some of the fishes kept in home aquariums. Swordtail minnows, bettas and paradise fishes may also change from males into females. A reverse change—from female to male—occurs among the black sea bass (*Centropristes striatus*). The young fish are predominantly females and produce normal eggs. When these fishes, which are common along the Atlantic Coast of the United States, reach five years, some change sex and become fertile males.

Functional hermaphrodites that can be both male and female at the same time were once thought to be extremely rare among fishes. But recent research has discovered that several large families of fishes include members that are simultaneously male and female. These include the family of the sea perches or sea basses (Serranidae) and that of the porgies or seabreams (Sparidae).

One of the most astonishing of these hermaphroditic sea perch is the small squibble sea perch (*Serranellus subligarius*), which lives in coral reefs off Florida. Each sexually mature fish carries both ripe eggs and sperm, and during the spawning season each fish may take male and female roles in turn.

Self-fertilization has been discovered in a small topminnow, *Rivulus marmoratus*. This tiny fish lives in brackish waters and low-lying tidal areas from Florida to the West Indies; it usually sheds its eggs, but some retain them in the ovaries. In an experiment, an egg from this fish kept in captivity was incubated and the young fish was reared on its own in an aquarium. The fish reached sexual maturity in six months and laid eggs. One egg was incubated and the young reared in isolation. It produced fertile eggs after four months. Thus this small fish produced self-fertilized eggs for two generations.

Organs of both sexes have been found in some deep-sea

fishes belonging to the order Iniomi. These include a lancet
fish (*Alepisaurus ferox*), a barracudina (*Lestidium pofi*) and six
bottom-dwelling species of tripod fish. We don't know defi-
nitely whether these fish are capable of cross-fertilization, but,
if it were so, it would be a useful means to ensure the survival
of the race. In the vast empty reaches of the deep seas,
meetings between the sexes may be infrequent. If individuals
were both able to lay and fertilize eggs, that would be a
decided advantage.

Marine fishes have definite seasons for spawning. These
are closely linked to the particular temperature of the
water—that is, most species have a preferred temperature
range that is best suited for spawning and the successful
development of the eggs. For most fish, too, there are not only
spawning seasons but also spawning grounds where they
assemble. The spawning grounds are places where there's food
for the young.

The migrations of some fishes rival those of birds, and
they move for much the same reasons—to escape cold, to
ensure food supplies and to breed. Their migrations cannot be
considered in detail here, but a typical instance is that in the
North Sea most fishes spawn during the spring or summer
when water temperatures are rising. Most freshwater fish also
spawn in the spring or summer, but salmon and trout spawn
from October to January in the Northern Hemisphere.

Like other eggs, most fish eggs, as you have already seen,
have a yolk, a storehouse of food on which the embryo can live
and grow during incubation or, in some instances, after
hatching.

Most of the fishes of the open seas lay floating (pelagic)
eggs. These are usually small translucent spheres that average
about four-tenths of an inch in diameter. The egg is enclosed
in a transparent membrane and often has a globule or two of
oil which assists its buoyancy. The eggs often have devices that

attach them to each other or to objects floating in the sea. For instance, the eggs of needlefish have small projections that become entangled with each other; those of flying fishes have long filaments that become caught in the sargassum weed or in debris. Most of the codfishes and flatfishes lay pelagic eggs.

Fishes that range the seas and lay pelagic eggs must lay them in even larger numbers than the others because the mortality of their eggs and larvae is particularly high; some die from bacterial and fungal diseases; others are eaten, especially by jellyfishes and arrowworms; and many eggs, particularly those that float, drift into waters that are unsuitable for their development. Therefore, the number of eggs laid in a season by a single female are astronomical. Marine biologists once counted the eggs in a 17-pound turbot and found it had nine million eggs. A female cod, depending on the species, lays between two million and nine million eggs at a single spawning. If all the turbot eggs were to survive and grow into adults, the seas would be thick with turbot, but, on average, less than one egg in every million survives to produce an adult and maintain the race.

Some studies of what happens to the pelagic eggs of the Atlantic mackerel (*Scomber scombrus*) show that the only hope of survival lies in prodigality. The Atlantic mackerel spawns in the western North Atlantic in coastal waters from Newfoundland to Chesapeake Bay. A survey in 1932, reported by Mr. N. B. Marshall in *The Life of Fishes*, found that by the time the tiny fishes were about two inches long, from one to ten fish only had survived from each million eggs. And this handful of survivors would have to face many other hazards before they would be ready to breed.

Let us follow the development of the fry of one of the codfishes, *Gadus morhua*, which is more or less typical of those fishes that shed pelagic eggs. *Gadus morhua* is distributed over both sides of the Atlantic. The mother of this particular tiny larva shed her eggs in the North Sea sometime during the period from January to April. Depending on her size, the female may lay from two million to nine million eggs. If this larva is like most of its fellows, the egg from which it hatched was shed in spawning grounds lying over banks to the north of Flamborough Head, England. First, as a tiny egg, it drifted with others of its kind. Some 10 to 14 days after it was shed in the seas, the larva, which measured a little over one-tenth of an inch, struggled out of the egg membrane and bore with it a supply of yolk to sustain it for three or four weeks. After about four weeks, the tiny fish began to live on zooplankton (microscopic animals). For as long as two and a half months the tiny cod fry may drift and grow with the plankton in the seas. Then, about midsummer when it is about three-quarters of an inch long and easily recognizable as a young cod, it moves down to the bottom in low inshore water and begins to feed on small crustaceans.

On the assumption that our particular small cod fry has survived all the perils of the deep, it continues to grow and to move out into deeper water. At the end of its first year of life it will have reached a length of about six inches. After three years it will be about a foot long and will be known to us as a codling. Our young cod will reach sexual maturity in its fourth or fifth year when it will have reached a length of two or three feet. Then, like its parents, it will be ready to spawn and start the cycle over again.

Other marine fishes lay nonbuoyant (demersal) eggs that sink to the bottom. Typical of these eggs are those of herrings, which are coated with a sticky substance and adhere to seaweed, stones and shells. Most of the marine fishes that lay the nonbuoyant, adhesive kind of eggs live and breed in offshore waters. These include gobies, glennies, sand eels, gunnels, bullheads, lumpsuckers, clingfish and toadfish. These fishes generally produce larger eggs than the floating eggs.

(Demersal eggs are also the rule among most freshwater fishes.)

A typical example of a marine fish that lays demersal eggs is the winter flounder (*Pseudopleuronectes americanus*) which inhabits the Atlantic Coast of North America from Northern Labrador to North Carolina and Georgia. The female lays from half a million to 1.5 million eggs which fall to the bottom and stick together in clusters. The eggs range in diameter from about three-tenths to four-tenths of an inch.

Over millions of years both kinds of marine fishes—those that lay pelagic and those that lay demersal eggs—have adapted to those conditions that best ensured the perpetuation of their species. For instance, demersal eggs are usually laid in areas that are also the nursery or feeding grounds of the larvae; these are sometimes in boisterous waters where floating larvae would be swept away from the nursery grounds. And the larvae of fishes that lay floating eggs, such as the codfishes, are generally those that feed on zooplankton.

10

EGG BURIERS

As we have seen, in the open seas most fishes lay floating eggs; in coastal waters many lay non-buoyant eggs; and in freshwater nonbuoyant eggs are the general rule.

In the course of evolution some freshwater fishes acquired the practice of burying their eggs. This provides some protection for the eggs and, thus, as a general rule, the egg buriers need lay fewer eggs than fishes which leave their eggs to chance.

The best-known egg buriers are salmon and trout. Brown trout (*Salmo trutta*) were introduced into North America many years ago and to other parts of the world. (Incidentally, there are no natural trout in the Southern Hemisphere, but brown trout and rainbow trout (*Salmo gairdnerii*), a native to western America, have been introduced widely.)

The behavior of the brown trout is typical of the Salmonidae family. The seagoing forms of this European fish move upstream in the spring. When spawning, the female digs a hole in a gravel bed (or redd, as it is called) with swift thrusts of her tail. Into this she sheds her eggs, which are then fertilized by a successful male suitor—that is, the one who has succeeded in driving away his rivals. The female lays up to 1000 orange-colored eggs for every pound of her weight. Brown trout grow up to 20 pounds or more. After spawning, the eggs are covered with gravel carried by the current. Fast-running water is essential so that the eggs are continually aerated during the long incubation period which can be from two to three months, depending on the temperature of the water. (Most trout and salmon prefer cold or temperate waters.)

When newly hatched, the larvae of brown trout have a large globular yolk sac attached and are about four-tenths of an inch in length. For three to four days they rest in the redd, then, as the sac acquires a more elongated shape, they burrow into the gravel. All this time they are developing the strength and jaws to catch their food of small insect larvae, small worms and crustaceans. Some weeks after hatching, when the yolk sac is completely absorbed, the fry emerge from the redds ready to look for food. Incidentally, while they've been in the redds, they have fanned their pectoral fins and flexed their tails constantly to prevent silt settling on them. This action has also stopped the silt from falling on some of the unhatched eggs.

Unlike the six species of Pacific salmon (*Oncorhynchus* spp.), trout don't die after spawning but continue to spawn for several years. Pacific salmon don't eat when on their spawning runs and grow steadily thinner and weaker. Some, indeed, never complete the return journey to the shallow freshwater streams where they were hatched, but perish on the way. Almost all Pacific salmon die after spawning. A few do survive

FIG. 4.   *Red salmon on their spawning beds.*

and return to the sea. Why the Pacific salmon and the
Atlantic salmon (*Salmo salar*) should make arduous journeys of
hundreds and even thousands of miles to return to ancestral
freshwater from the oceans, leaping as high as ten feet up
waterfalls, is a mystery. The generally accepted theory is that
salmon developed originally in freshwater and only later
acquired the practice of going to sea in search of food.

How they find their way, generally to the very stream
where they were spawned (see Figure 4) is another mystery.
One theory is that they follow currents. Another is that they
steer by the sun or the stars. Yet another is that they have a
built-in directional memory and recognize their home stream
by its smell.

In 1951 two American ichthyologists, A. D. Hasler and
W. Wisby, tested the "smell theory" by capturing some
spawning coho or silver salmon (*Oncorhynchus kisutch*) from two
branches of the Issquah River in Washington. They plugged

the nasal sacs of half the captured fish with cotton batting
before releasing them. The fishes without the nasal plugs were
able to find the stream from which they had been taken, and
most returned to it. But the plugged fishes were not able to
discriminate and pursued a random pattern.

But even without nasal plugs, salmon don't invariably
return to the streams where they were hatched. Experiments
have shown that 14.9 percent of silver salmon fingerlings
hatched and tagged in a California stream failed to return to
their home waters and strayed into an adjacent one four and a
half miles away.

Some experts think that salmon may have little directional sense at all, that they swim at random on their return to
the coast, and that because of their immense numbers enough
will always return to their parent river.

Only extensive research can help to answer this question

and, equally interesting, the question of where the young
salmon go when they go to sea. Some light on where larger
Atlantic salmon feed at sea was shed in the late fifties when a
U.S. nuclear submarine, cruising under the ice in the David
Strait between Baffinland and Greenland, saw thousands of
salmon hanging like silver icicles from the underside of the ice
pack.

Trout and salmon bury their eggs. So do their relatives
the grayling (*Thymallus* spp.) and char (*Salvelinus* spp.). The
brook trout of eastern North America (*Salvelinus fontinalis*) is a
char. This is a handsome fish with a mottled pattern usually of
red spots on the side.

Other fishes that spawn in gravel or sand include the
European barbel (*Barbus barbus*) and certain North American
darters, such as *Percina caprodes*, *Etheostoma caeruleum* and *E.
spectabile*. Because the eggs are comparatively safe from predators, all these fish need lay fewer eggs than do their cousins in
the ocean, and they also lay larger eggs so that larvae are
bigger when hatched.

Other egg buriers are the so-called annual fishes whose
life-span rarely, if ever, exceeds a single year. These fish, which
are small topminnows, live in the drier parts of South America
and Africa where the short rainy season is followed by a hot,
dry period during which most of the streams dry up. Before the
water evaporates, the small topminnows (belonging to genera
such as *Aphyosemion* and *Cynolebias*) bury their tough-coated
eggs in the mud. They don't begin to incubate until the
annual rains come again. The young fishes have plenty of
food, often, in fact, minute organisms which have fed on the
decomposed bodies of parent fish. The incubation period
varies from species to species, from a few weeks to some
months. Among the more beautiful of these annual fishes are
the African lyretails (*Aphyosemion* spp.).

It is thought that the presence of certain bacteria is

necessary to precipitate the hatching. Aquarium keepers have triggered the hatching of one of the African lyretails, *Aphyosemion arnoldi*, by adding a little powdered milk to the water. Dealers in some of these tropical annual fishes send the eggs by mail, packed in peat moss, just before they are about to hatch. The receiver has merely to immerse the peat moss in an aquarium tank and he'll have fry swimming in the water in a few hours.

You may be interested to compare the means of survival of the annual fishes and the lungfishes which we looked at earlier and which live in a similar habitat where a rainy season is followed by a hot dry one. The young lungfishes survive by burying themselves in the mud and waiting for the rains. The annual fishes die but perpetuate their species by burying their eggs in the mud.

Burying the eggs does provide some degree of protection —much more than if they were left to float or sink to the bottom. In the following chapters we'll see how other fishes provide other forms of protection for their eggs.

11

LIVE-BEARERS

WHEN HE NETTED some small, beautiful fish from a stream on the island of Trinidad in 1866, little did the Reverend Robert John Lechmere Guppy realize that he was going to achieve immortality and also give a lot of enjoyment to owners of home aquariums. The little two-inch fish with the streamerlike tails are *Lebistes reticulatus* or, more familiarly, guppies, the best-known of the live-bearers.

They're prolific breeders, too, as you probably know if you've ever kept them. An average female produces 20 to 25 young every four to six weeks. A large female may produce 60 to 70, and she may do this for several months after a single internal fertilization; the frequency depends on factors such as the temperature of the water. This high fertility once earned the guppy the name of "the million fish." (In the usual aquarium the water doesn't become thick with guppies because other fishes, including their own parents, eat them. If you want to breed guppies, you have to put a pregnant female in a tank on her own and remove her after her young are born.

Most females you buy are pregnant—you can tell if she is by looking for a dark patch [gravid spot] showing through the flesh at the end of the body cavity.)

The olive-colored female of the wilds is about twice as long as the bright-hued male. (Colors are rarely the same in two individuals.) You may be lucky enough to see a number of gay-hued males courting a female with swimming dances, looking, as someone has said, much like tiny hummingbirds flickering around a blossom. After the courtship, one of the males fertilizes the female by means of his *gonopodium,* an elongated anal fin which transfers the sperm into the female's body. This copulatory fin is longer than the other anal fins and allows you to identify instantly any male of the family Poeciliidae (poeciliids) which includes the familiar mollies, mosquito fishes, swordtails, platys and other topminnows, as they're called. All told there are 45 genera and a great number of species in this New World–restricted family, ranging from Illinois and New Jersey in the United States to Argentina. The guppy is found in Venezuela as well as Trinidad.

Lebistes reticulatus is an ovoviviparous species—the fertilized eggs develop inside the mother, fed mainly by the yolk. It is thought that guppy eggs are also fed by a secretion in the mother's sac of cells. A larger member of the poeciliids, the green swordtail (*Xiphophorus helleri*) that grows to three inches, is more fertile than the smaller guppies, dropping up to 200 young at a time.

Another member of the family, *Heterandria formosa,* one of the smallest of the mosquito-eating fishes, is a viviparous species—the young are nourished by a placentalike arrangement of blood capillaries inside the mother's body. Like *Lebistes,* the one-inch-long female is able to store sperm and give birth to successive broods over some months after a single fertilization.

The range of live-bearing fish is considerable, from those

mothers who provide little more than a sanctuary for the developing eggs (the ovoviviparous species) to those viviparous species like *Heterandria* just described. You will remember that some of the live-bearing sharks carry their young for many months.

Another live-bearer, the most famous mosquito-eating fish of all, *Gambusia affinis,* also belongs to this family. From its southeastern United States habitat, this two- to three-inch fish has been exported to many parts of the world and released in swamps to control mosquitoes by feeding on the larvae.

Most wild specimens of *Mollienesia latipinna,* the broad-finned molly, are gray with brilliant metallic reflections and rows of dark spots on the dorsal and caudal fins. But occasionally a black or *melanistic* one is born. In the wilds its life is usually short—it is too conspicuous and is more easily seen by predators. From the black mollies, breeders have developed a black race.

All the poeciliids give birth to miniature replicas of adults. So do two other families of topminnows, the Janynsiidae and Goodeidae.

The family of four-eyed fishes, Anablepidae, are also live-bearers, nourishing their young on a kind of placenta and producing fully formed young. The most famous four-eyed fish, able to see well both in air and water, is, of course, the archerfish which shoots down insects with drops of water.

Other live-bearers include the halfbeaks (Hemiramphidae) of both salt and freshwater in India and Asia and the sea perches (Embiotocidae) of the North Pacific. These perchlike fishes usually live close to the coasts. The fertilized eggs are nourished by secretions from the mother's ovaries. She gives birth to from three to eighty, and an unusual feature is that males may stay in the ovary until they are sexually mature. Thus they may mate soon after birth.

The female eelpout (*Zoarces viviparus*) of Europe gives

birth to from 20 to 300 young, depending on the size of the mother.

There are also some live-bearing members of the scorpion-fish family. The rosefish (*Sebastes marinus*) is one of the family found on both sides of the Atlantic. One 13-inch female was found to carry 20,000 young.

The bony fishes thus have evolved various ways of reproducing themselves. Most lay eggs which are externally fertilized, and they must lay them in immense numbers. A few fishes have developed ways of nourishing their young in their ovaries and produce live young which are often miniatures of their parents. In some the eggs are merely incubated in the ovary. In general, the live-bearers need to produce fewer young than those fishes that lay eggs and leave them to chance. The same holds true of those fishes that take some care of their eggs and young—which are the subject of the next chapter.

12

MOUTHBREEDERS

SOME CATFISH FATHERS of tropical and subtropical regions are among the most "devoted" parents of the wild and are, of course, among the few exceptions to the general rule of leaving eggs—and offspring—to chance. I say "devoted" because fishes don't have human emotions. Nonetheless, if we saw a human father behaving with as much solicitude we'd call it devotion. These baby-sitting catfish fathers belong to the ariid family (Ariidae) which are world-wide in distribution. One member known on the American coast from Cape Cod to Panama is *Galeichthys felis.* This small catfish breeds in June or July. After fertilization, the male picks up in his mouth from 40 to 55 eggs and carries them there for a month until they hatch. Then he holds his youngsters there for another two weeks. During all that time he goes without eating.

After they hatch, their father's mouth is a sanctuary to which they hasten if they feel threatened. If they're laggard, the father gathers them up.

All told, about 40 species of ariid catfishes incubate their young. An American naturalist, Jean George, has described in a *Reader's Digest* article how one day she watched a school of 27 tiny catfish with big heads and black whiskers swimming round in a pond, watched over by their father. Her dog apparently frightened the tiny fishes, which scattered in 27 directions. It took the father one hour to gather his family together.

Some other catfishes, notably the bullheads (Ictaluridae) of North America, don't incubate their young but build nests and guard the young during the first few difficult days of life. Nest-building fishes are discussed in the next chapter.

Most of the ariid catfishes are marine species and they usually breed in saltwater. But breeding can take place in freshwater, depending on the species.

Mr. Earl S. Herald, a marine biologist, describes in *Living Fishes of the World* his surprise at the size of the eggs of one big ariid catfish, *Arius dispar,* which he picked up by the tail in the Philippines. In this large species they were seven-eighths of an inch—bigger than some marbles.

Other tropical and subtropical catfishes have other equally astonishing methods of brooding their eggs. Some male members of the Loricariidae family of armored catfishes carry their eggs in folds of skin near the lips. Then there's the female of the bunocephalid catfishes (*Aspredo* and *Bunocephalus*). She rolls on the eggs after they have been fertilized so that they sink into the soft, spongy skin under her body, and the lower part of her head and abdomen become covered with eggs. After a time each egg is borne on a stalked cup growing from the skin, which, supplied with blood vessels, feeds the embryos. Not surprisingly, these fishes are popularly known as obstetrical catfishes.

Some marine cardinal fishes also incubate their eggs in their mouths. Probably the record for the number of eggs incubated in the mouth is held by the males of one of the Mediterranean species, *Apogon imberbis;* one male may hold up to 20,000 eggs in his mouth—each about two hundredths of an inch in diameter. He swills the eggs continually to keep them sufficiently aerated.

Just as some species of marine fishes do, certain fresh-water fishes brood their eggs in their pharynx and mouth. These include some cichlids (Cichlidae), familiar to those who keep tropical fish. There are over 600 species of these spiny-rayed, perchlike fishes, distributed widely in South America, northward to Texas, and in Africa.

More or less typical of the mouthbreeding cichlids is an African member of the large genus *Tilapia, T. mossambica,* who incubates her eggs for 10 to 12 days, the whole time moving them about so that they are aerated. (If you take her eggs and try to incubate them, you'll usually fail; they grow moldy and die.) When the 20 to 25 eggs hatch, the young school in a cloud around her head and follow her about. Innate behavior causes them to seek her mouth, eyes and the dark shadows made by her two pairs of fins. When she feels her young are threatened, she opens her mouth. Like a black cloud the school eddies into her mouth until the danger is past. After a week or so the young disperse to fend for themselves.

T. mossambica are famous as a rich source of protein in Asia and East Africa where they're cultivated in ponds. Fish-farming research at the Tropical Fish Culture Research Institute at Malacca has shown that yields can be stepped up to a ton of fish per acre per year by adding 300 pounds of superphosphate to the water. The natural yield is 20 pounds of fish per acre per year. The superphosphate feeds the growth of plants on which small fishes feed; they in turn provide food for the *Tilapia.* The return of one ton of protein per acre per year is incomparably greater than you'd get from a similar expenditure of 300 pounds of superphosphate to grow grass for sheep or cattle.

In Java the peasants think this wonder fish a gift to them from the gods. And no one can blame them; in 1939, some peasants found five *T. mossambica* in a rice field, thousands of miles from the fish's African home. How they got there is an unsolved mystery.

The wonder fish, which is also extensively farmed in its East African home, multiplied rapidly in the Javanese rice fields and has proved to be the means of saving millions of rice-eating Asians from dangerous diet deficiencies. In a joint effort, the United Nations and the United States have in the postwar years been developing large-scale farming of *T. mossambica* in the Middle East and Southeast Asia.

The female *T. mossambica* broods the eggs in her mouth, but in other species of this genus the male picks up the eggs and incubates them. Mouthbreeding is usually carried out by one parent, but among another cichlids, the South American discus fish (*Symphysodon discus*) the task is shared by both. Both parents can be said to "suckle" their young. This lovely freshwater fish, as beautiful as some of the tropical marine butterfly fishes, is probably the most remarkable of the cichlids. Both parents take turn in guarding, fanning and mouthing the eggs. When the larvae emerge, the parents transfer them to water plants and rocks where the young remain, wriggling at the end of a short thread exuded from the head. For four days the parents stand guard over the tethered youngsters; then the larvae break free, swim to their parents' sides and feed, it is thought, on nutritious secretions from the parents' skins. The parents even share this "suckling." With a flick of its body, one parent will transfer the whole cluster of up to 200 tiny fishes to the other parent. When not feeding, the young stay close to their parents. The nursing of the young fishes lasts for at least five weeks.

T. mossambica is described in detail because in most respects it is typical. Another cichlid mouthbreeder, *Haplochro-*

mis multicolor, of Africa, is remarkable because the olive and silvery-colored male bears four unusual spots on his fins. The spots closely resemble the eggs that the female lays and later gathers in her mouth to brood them. Why do the spots so closely resemble the eggs? The answer came when observers noted during spawning that the female nibbled the "eggs" and tried to take them into her mouth. When she thus approached the male so closely she took water and milt into her mouth and this fertilized the eggs she had gathered. Truly a remarkable piece of deceit acquired over millions of years of evolution to aid fertilization.

Why don't cichlid parents eat their young? There's some evidence for believing that cichlid parents learn not to eat their young. Biologists removed the eggs from a pair of cichlids, breeding for the first time, and replaced them with the eggs of another species. The cichlid couple successfully raised the "cuckoo" brood, which they thought was their own. But later, when they were allowed to breed, they ate their own young as soon as they hatched. The probable explanation is that the parents were "imprinted" with the first brood and, therefore, did not respond with parental instincts to the second—and different—brood.

Mouthbreeding does give the eggs and babies a better chance of survival. Altogether there are 200 species of mouthbreeders in predator-infested waters throughout the world.

Gill chambers are also used by some fishes for incubating their eggs. Notable instances are two species of North American blind cave fishes (*Amblyopsis*). The female broods some 70 eggs in her gill chamber. One captured fish had 70 eggs that took two months to hatch and produce miniature replicas of herself, three-eighths of an inch long.

The male kurtus (*Kurtus* spp.), a perchlike fish of fresh and brackish water in the Indo-Australian region, incubates its eggs, not in its mouth, but suspended from a bony hook on its forehead. The eggs are usually laid in two bunches connected by a fibrous thread. It is thought the male detaches the eggs with this hook from aquatic plants where they've settled.

A mouth is not the best place to brood eggs. Some tube-mouthed fishes have evolved special pouches for use as brood chambers. Female ghost pipefishes (*Solenostomidae*) have semiopen brood pouches formed by the enlarged pelvic fins. After fertilization, they transfer the eggs to these pouches, where they are secured by short filaments projecting from the inside.

The true pipefishes and the sea horses are much more famous because, in popular terms, they have been called the "henpecked fathers of the seas." The males care for the eggs. Some pipefishes merely attach the eggs to the undersides of their trunks or tails, but other species and the sea horses have special pouches.

All the tube-mouthed fishes live by siphoning their food. They aren't fast enough swimmers to catch prey, so they have evolved a deadly suction method to gather, usually, minute crustaceans into their toothless mouths. Someone once described pipefish as resembling pipe cleaners that had come to life—accurate enough if you realize that the ridges in the living "pipe cleaner" are concentric rings of bony plates.

Sea horses have been renowned—and named—ever since Pliny the Elder (23–79 A.D.) called one *hippus* or "horse." It's retained today in the typical generic name *Hippocampus*, loosely "horse that bends," referring both to the head and the prehensile (grasping) tail. There are about 24 species, distributed throughout the world and ranging from the pygmy sea horse (*Hippocampus zosterae*) of the Gulf of Mexico, only one and a half inches long, to the common Atlantic sea horse (*Hippocampus hudsonius*), which reaches about five and a half inches.

A great deal of folklore has accumulated around these strange creatures. Their flesh has been used in love potions, and, in powder form, they were said to be a cure for baldness.

It is, indeed, an odd-looking creature. In the May–June, 1968, issue of *Pacific Discovery*, the naturalist Mildred D. Bellomy described it as having, in miniature, a tail like a monkey, a body like an insect, a head like a horse, eyes like a chameleon and armor like King Arthur's knights.

The eyes do resemble the chameleon's in that they work quite independently of each other. One eye may be looking at some tasty morsel of brine shrimp lying ahead, and the other looking upwards warily for danger or (in the breeding season) for a mate.

They swim vertically—I was tempted to say "drift"—and more often than not, artists show them incorrectly with their tails coiled backwards towards the back of the head. Young sea horses may bend their tails that way, but adults never do. The tails can only coil round vegetation when coiled forward towards the snout.

Sea horses vary considerably in color. For instance, the common Atlantic American sea horse may vary from pale white through deep red to blue-black.

Courtship varies in detail from species to species. In one species, a courting pair intertwine their tails in a kind of lovers' knot; in another, the males pump water out of their pouches and look as though they are gravely bowing to the females, much like courtiers performing an underwater ga-votte! Courtship may extend over a day or two, during which time the male and female, sometimes with clasped tails, sway towards each other rhythmically. Then one of the pair begins quivering from tailtip to coronet-topped head.

Courtship ends when the female inserts her bony egg-laying organ (ovipositor) in the male's pouch, which lies just above the tail. She deposits her eggs until the pouch is filled.

Then she leaves and, if she has surplus eggs, seeks another male. Or sometimes one female hasn't enough eggs to fill the pouch, and two and even three females may be needed to fill it. Some sea horse fathers, such as the Mediterranean short-nosed sea horse (*H. brevirostris*), may brood up to 200 eggs.

The size of the eggs varies from species to species—from one-fiftieth of an inch to a quarter of an inch in diameter. Some are a beautiful salmon-pink color, enclosed in a transparent shell.

The brood pouch, a spongy, quadrangular network, is provided with square-shaped depressions for each egg and is richly supplied with blood vessels, which convey oxygen and a nutrient secretion to the developing embryos. This placenta-like lining develops before mating.

The brooding time varies, depending on the species, and even within any one species, depending on the water temperature. The average period of gestation is about three weeks. Then the male enters "labor," expelling the tiny babies with vigorous contractions of his swollen pouch. This may take several days, with up to 30 babies being born more or less together.

Parenthood is almost continuous for the smallest of all the sea horses, *H. zosterae*, which lives off eastern Florida and in the Gulf of Mexico. This one-and-a-half-inch sea horse breeds about nine months of the year. Two days after giving birth to about 50 babies, he is ready to brood another lot.

Baby sea horses are miniature replicas of their parents, almost transparent, but ready to fend for themselves. Pigmentation develops later. Most species probably aren't ready to mate until they are about a year old. But the smallest of all, *H. zosterae*, is sexually mature when only two or three months old, and its life-span is about a year.

The eggs of all the fishes we've looked at in this chapter

have been given some protection. Thus, these fishes lay far fewer eggs than those which leave their eggs after spawning.

Other fishes have developed other ways of guarding their eggs. Some, for instance, build nests. They're the subject of the next chapter.

13

NEST BUILDERS

SOME FISHES SAFEGUARD their eggs by building nests. If you've kept some of the labyrinth fishes, such as gouramis, you may have seen the male gourami building his air-bubble nest. He does so by blowing a mass of bubbles at the surface of the water so as to form a nest in which the female lays her eggs. Why do the bubbles stick together in a thick mass? Before blowing each bubble, the male coats it with a sticky mucous substance.

These fishes are called "labyrinth" fishes because they can breathe air through a labyrinthine organ in each gill chamber. Ichthyologists refer to them as the anabantids. They include the Siamese fighting fish (*Betta splendens*), the croaking gourami (*Trichopsis vittatus*), the thick-lipped gourami (*Colisa labiosa*), the dwarf gourami (*Colisa lalia*) and the paradise fish (*Macropodus opercularis*).

This organ allows these fishes to survive in stagnant water that is short of oxygen—much as the lungfishes do. The adaptation arose just as it did with the lungfishes—in condi-

103

tions where seasonal rains were followed by long periods of drought.

Only by building bubble nests can these fishes ensure sufficient oxygen for their eggs in still and stagnant water in the tropics. Here, again, is an adaptation that has aided in survival.

Air-bubble nests for breeding are also built by some air-breathing armored catfishes (*Callichthys callichthys* and *Hoplosternum* spp.) which also live in stagnant water.

As well as building nests, some male labyrinth fishes also stand guard over the eggs while they are hatching. Nest-building fishes are thus defenders of territory, much as birds are. (So, too, are some fishes that don't build nests, such as some of the brightly colored inhabitants of the world's coral reefs. The bolder the colors, the more property-conscious is the usual rule.)

No male labyrinth fish defends his territory more fiercely than the well-named Siamese fighting fish, so familiar to all of us. The nest-watching and breeding routines of this fish are typical of the group, but the ones we see in aquariums are much more handsome than their wild relatives. Selective breeding over many years has developed a rich range of color—brilliant deep blue, light blue, or blood red—and enlargement of the fish's asymmetrical dorsal, anal and tail fins. The wild fish is much less glamorous, brownish-yellow to cream in color with a few darker stripes.

In Thailand, contests (now illegal) were held between specially bred fishes. Flushing with color, the two pugnacious little fishes hurled themselves at each other, using their snouts as battering rams. Tail fins were torn first, then anal fins as the two charged each other repeatedly. Blood began to flow from the fins and the fishes' sides. The battle would often last an hour before one fish withdrew.

Fights between wild fishes do not last nearly so long—the

fish which intrudes on a nest guarded by an aggressive owner usually retreats after a few minutes.

The male Siamese fighting fish usually builds a little floating basket of mucous-covered bubbles in the course of a single night of strenuous work. The first few bubbles are usually attached to a floating leaf or stick. The circular nest may be about three inches in diameter and almost an inch thick, and he defends it vigorously against any male who approaches. Meanwhile, he displays himself to females, spreading his handsome fins and flushing with color.

Some females aren't ready, but eventually one is impressed and lingers. The male makes little plunging dances, pivoting about her. The courtship culminates in a wild chase through the water, at the end of which he persuades her to place herself under the nest. He then wraps his body around hers and appears to squeeze the eggs from her body. (This may not be so; his actions may merely trigger a reflex in the female.) According to the female's age and condition, she may lay 10 to 50 little whitish globules, which the male fertilizes as they sink to the bottom. From there he retrieves them with his mouth and spits them into the nest.

Then he takes up his vigil, driving away all other fish, including the female. Should an egg slip to the bottom, he retrieves it. He may also repair the nest by blowing more bubbles into it. The male may court several females.

After a few days the larvae hatch, and they, too, may be gathered up and spat back into the nest, should they leave it. The male's vigil continues until the fry are about six days old.

Some cichlids excavate a nesting pit on the bottom, and the eggs are guarded by a parent. Sometimes the fry, too, are guarded for a few days by the parent. Konrad Lorenz, in *King Solomon's Ring*, gives a graphic account of a cichlid parent carefully putting his children to bed at night. As night nears, a cloud of small fish will follow the parent to the pit. But not all;

many loiter outside, and these he takes in his mouth and spits into the nest. You might think it a hopeless task—as soon as the father puts one in the nest, others swim out. But nothing of the sort happens. Says Lorenz:

> The baby sinks at once heavily to the bottom and remains lying there. By an ingenious arrangement of reflexes, the swim bladders of the young "sleeping" cichlids contract so strongly that the tiny fishes become much heavier than water and remain, like little stones, lying in the hollow, just as they did in their earliest childhood before their swim bladder was filled with gas. The same reaction of "becoming heavy" is also elicited when a parent fish takes a young one in its mouth. Without this reflex mechanism it would be impossible for the father, when he gathers up his children in the evening, to keep them together.*

Some North American catfish, the bullheads, as we saw in Chapter 12, also make nests and guard the eggs. The common bullhead or horned pout of New England lays some thousands of eggs in a hole made by the male. (Sometimes he just takes over a rodent's abandoned tunnel.) The male's protection extends to the fry; he guards the school until the fry are about two inches long.

Common names of freshwater fishes are often confusing. There are fish called "bullheads" in Europe, which also build nests and guard their eggs but aren't catfishes; they're small sculpins of the Cottidae family. The best-known is the four-inch-long bullhead or miller's thumb which is found in swift, cool streams in most of Europe. The female lays her clusters of 200 to 300 pink eggs in the nest, and the male takes up guard. During the three to four weeks the eggs take to hatch, the male fans the eggs with his pectoral fins to prevent them being covered with silt. If this were to happen, the eggs would die.

A marine relative of the bullhead deserves our attention

---

* Konrad Lorenz, *King Solomon's Ring* (London: Methuen & Co., 1952).

because the female reverses the usual order of things and pursues the male. The female grunt sculpin (*Rhamphocottus richardsoni*) chases the male until she corners him in a small crevice or cavern in the sea, and keeps him there until he fertilizes her 150 or so eggs which she lays on the walls of the prison. Grunt sculpins are so called because of the noise they make when taken from the water. They're marine fishes of the American Pacific coast.

One of the most celebrated of the nest builders and nest defenders are the sticklebacks. They're favorites with home aquarium owners because of their fascinating breeding behavior and, moreover, because they have been much written about by two eminent naturalists, Konrad Lorenz and Niko Tinbergen.

There are several stickleback species native to Europe and North America, distinguished from one another by the number of spines. The three-spined stickleback (*Gasterosteus aculeatus*) is probably the most cosmopolitan of the family. It's found on both sides of the Atlantic and in both fresh and salt water.

The mating behavior of this tiny, aggressive fish is typical. In autumn and winter the three-spined sticklebacks swim in small, mixed schools, but in early spring each male leaves the school and lays claim to a territory, much as a songbird does. Any other stickleback, male or female, is attacked if it intrudes.

Then the male builds a nest. He digs a shallow pit in the bottom by carrying away the sand mouthful by mouthful. When he has formed a depression about two inches square, he starts on the construction of the nest; he collects pieces of waterweeds, "cements" them together with a sticky secretion from his kidneys and shapes the whole into a mound with his snout. When sufficient material is accumulated, he hollows it into a tube by boring through it.

The nest ready, the male undergoes a spectacular change of color. Normally an inconspicuous gray, he now acquires a bright-red chin and undersurface and a shining bluish-white back.

He defends his nest vigorously, staying close by—and as a consequence many male sticklebacks are taken by cormorants and herons.

Meanwhile, changes have also been taking place among the females, which have become silvery and swollen with 50 to 100 large eggs. Should a female, silvery and heavy with eggs, swim by in a school, the male woos her by swimming a leaping, zigzag figure round her. If she is ready, she swims towards the male with her head held high—a signal which the male understands. He turns and swims swiftly toward his nest, followed by the female. Arrived at the nest, he thrusts his head rapidly into it several times and displays his erected dorsal spines toward his mate. Then she enters the nest and rests there with both her head and tail protruding. The male prods the base of her tail several times with his snout. These prods stimulate her into laying a number of eggs. This done, she slides out of the nest, and the male takes her place to fertilize the eggs. Then he chases the female away and goes looking for another bride. He may persuade up to five females to deposit eggs in his nest.

Nesting behavior of a remarkable kind is shown by the bitterling (*Rhodeus amarus*), a small freshwater fish of central Europe. During the spawning season the female develops a long tube which she uses to lay her eggs in the gills of a freshwater mussel. The male has meanwhile been driving rival males away from the shell (his territory). He has acquired a brilliant courtship dress of reddish fins, orange-yellow underparts and a darker back. When the female is laying her eggs, he releases his sperm into the water nearby. The sperm is drawn into the mussel when it feeds by pumping in water.

The little pumpkinseed or common sunfish (*Lepomis gibbosus*) of North America is another nest builder, like others of this North American freshwater family. The male's behavior much resembles that of the sticklebacks. In late spring he acquires brighter colors and scoops out with his tail a hollow in the sand close to the shore. Passing females are wooed in turn and induced to lay in his nest, and the eggs are guarded pugnaciously. The fry, too, are guarded for a few days, after which they wisely disperse because otherwise he'd eat them.

Some members of the large family of blennies and their close relatives, such as pricklebacks and gunnels, also make nests and guard their eggs.

The female of one of the blennies, *Blennius ocellaris*, will sometimes lay her eggs in a sunken bottle and then guard the eggs by backing into the bottle's neck. In *Life in the Seas*, the English naturalist John Croft says that he once carried a milk bottle from the seabed off Wales to the surface with a female blenny inside guarding her eggs. The blenny made no attempt to leave until the surface was reached.

Nest defenders, as we observed earlier, are territory defenders. Battles between males are rarely mortal; they are decided by means of bluff, using signals. These signals aren't always visual; some fishes (cod, toadfishes, drumfishes and others) use sounds to intimidate and also for courtship. (Wooing by sound is the subject of Chapter 16.)

In this chapter we have looked at some of the more interesting of the fishes that build nests. Nest builders are the exception rather than the rule among fishes, as has been pointed out earlier; most fishes lay their eggs and leave them. Nest builders, and particularly those which guard the spawn, therefore need to lay fewer eggs to ensure survival.

Fishes cannot go much further to ensure the safety of their eggs—but some do. In the next chapter we'll look at some fishes who try to find security for their eggs by laying them on land.

14

FISH EGGS OUT OF WATER

AS WE HAVE SEEN, some fishes seek safety for their eggs by a variety of means—by keeping the eggs in their mouths, burying them, building nests and so forth. They couldn't do much more to guard them from snapping jaws, short of laying them on land. And this, indeed, is what some remarkable fishes do.

For instance, there's the characin, *Copeina arnoldi*, of the river Amazon and its tributaries. After the three-inch bluish-gray male has found a willing female, he steers her to a place where a bunch of leaves hangs an inch or two above the water. Then, locking fins together, both leap out of the water and cling to the vegetation for a few seconds—time enough for the female to deposit a cluster of eggs and for the male to fertilize them. They repeat this for some minutes until finally a largish mass of eggs has been deposited there.

The eggs are thus safe from waterborne jaws, but they face the hazard of being desiccated by the sun. The male prevents this by repeatedly splashing the eggs with water throughout the three days the eggs take to hatch. When the larvae do emerge, they drop into the water and swim away.

Equally astonishing are the grunion of southern-California waters which come ashore in sheets of shimmering silver to lay their eggs. These seven-inch-long fishes, *Leuresthes tenuis*, have a breeding season which lasts from late February to early September—a long time—but breeding takes place on a few nights only. These are the three or four nights that immediately follow the highest tides, when there's a new or a full moon.

There's a marvelous precision in all this. If a female grunion could think and speak as we do, she might put it this way: "I want to lay my eggs so that they will be protected by sand for ten days while they are incubating. If I lay them before the peak tides, they'll be washed away. Therefore, I must lay just after the peak tides following a new or full moon. And if I lay them on the night of the highest tide and on the following night, then the next high tide might not reach them."

A female grunion may spawn up to seven times during the year. Each time she lays her eggs when out of the water and unable to breathe, an astonishing instance of the reproductive instinct driving these little fishes to do what is contrary to normal behavior—to strand themselves high and dry.

Each female lays an orange-colored cluster of 1000 to 3000 eggs—far fewer than most fishes lay—and each tiny egg is about one-fourteenth of an inch across. Later and smaller tides pile sand on the eggs to a depth of six inches, or even a foot. Then, safe from the jaws of fish and the beaks of birds, the eggs incubate. The baby fish are ready to hatch in about ten days but will not do so until four days later when the next high tide exposes the eggs. (Apparently no grunion egg will hatch until it has been uncovered and rolled in the surf. In aquariums it has been found the eggs won't hatch until they have been rolled.)

FIG. 5. *Male grunion surrounding a single female at the edge of the surf to fertilize her eggs.*

Spawning starts when the grunion move in with the surf and leap onto the beach. It may begin when a wave falls back to leave one or more stranded fish. Stranded on the beach, the females arch their bodies and burrow violently with their tails into the sand until sometimes only their heads and pectoral fins are clear. They do this in about ten seconds. Males (who usually outnumber females two or three to one) wriggle around them and discharge their milt, which runs down the bodies of the females to fertilize the eggs that are now being deposited in the hole. From time to time a male wraps himself around a burrowing female. (See Figure 5.)

While the females are spawning they emit sounds rather like the squeak of a mouse. The name "grunion" is thought to be derived from the Spanish *gruñón* which means "grunter."

Their task performed, the males and females wriggle over the exposed sand to reach the security of the surf.

This species of grunion, because it spawns at night, is relatively safe from predators such as birds. But another grunionlike fish of the Gulf of California, *L. sardina*, often spawns during the day, just after the high spring tides, and is preyed upon by large flocks of seabirds.

Grunion belonging to the silversides family Atherinidae have a delicate flavor and in the past were gathered in truckloads from California beaches—so much so that biologists feared for the grunion's future. Today, licensed fishermen—and children under 16—may catch them in unlimited numbers—but only with bare hands—during March, June, July and August. None can be gathered during the peak spawning months of April and May.

A lot has yet to be learned about the habits of grunion, but apparently they do not venture far from the beaches of their birth during their three or four years of life. They grow to about five inches during their first year and are then ready to spawn. Males usually don't grow longer than six inches; females generally grow a little longer.

How does the grunion "know" when to spawn—what triggers the spawning runs? What sets off the extremely accurate "internal clock" that makes them beach themselves at the right time?

Possibly they can sense variations in the force of gravity, which produce the high tides and the changes in water pressure. Or it could be that they somehow measure and respond to the changes in the intensity of the moon's light.

It's a mystery that only close observation can unravel.

15

MYSTERY OF EELS

FRESHWATER EELS, declared ancient naturalists from
Aristotle to Izaak Walton, reproduced themselves
by spontaneous generation out of mud. Or, they speculated
later, they grew from a type of worm such as the intestinal
ones of the black goby. In Sardinia, they claimed, eels were
propagated by a species of black beetle.

It's all very well for us to scoff, but until the end of the
eighteenth century no one had ever seen an eel with mature
sex glands, and until 1897 no one had identified the larvae of
eels; and, further, their spawning site in the Sargasso Sea was
not discovered until 1922.

The tracking down of the spawning grounds is one of the
great detective stories of marine biology. It starts in 1856 when
a German naturalist, Johann Jakob Kaup, described a small,
translucent fish sent to him from the Straits of Messina. He
called this fish, which resembled a willow leaf with its veinings,
Leptocephalus brevirostris. The fish remained catalogued but
largely forgotten for 40 years until two Italian oceanographers,

Grassi and Calandruccio, took a closer look at it. Giovanni Grassi and Salvatore Calandruccio concluded that *L. breviros-tris* was a deep-sea fish carried up from the depths by the tides in the Straits of Messina. Then in 1897 they caught some little willow-leaf fishes that were even slimmer than *L. brevirostris* but equally transparent. They appeared to be turning into eels, and Grassi and Calandruccio realized that *L. brevirostris* must be the larva form of the freshwater eel (Anguillidae).

Grassi and Calandruccio were now able to reach some conclusions about freshwater eels and their mysterious breeding. The first conclusion was that eels presumably made their way from freshwater to the seas to breed. There had been observations of males making their way to the sea at four to eight years of age and females at about twelve years. The second was that eels must descend to the depths to breed. A number of mature eels had been captured with enlarged eyes—deep-sea fishes have large eyes in order to capture as much light as possible. These eels, moreover, had shown some slight development of their sexual glands. But where did the eels breed? No breeding ground was known in the Mediterranean or elsewhere.

The site remained unknown for many years more. In the first years of this century, Danish research ships searching for young cod and eggs in the Atlantic dredged up *Leptochephalus* larvae and sparked a line of investigation by a young Danish marine biologist, Johannes Schmidt. If you caught enough *Leptocephalus* larvae, it might be possible to track them as they became progressively smaller. He began his search with fine meshed nets towed at varying depths over the North Atlantic.

Johannes Schmidt spent 16 years at sea in his famous ships the *Thor* and the *Dana*. Year after year the search was narrowed down. At the start of World War I, Schmidt had this picture: To the west of Great Britain, eel larvae, *Anguilla anguilla*, averaged 75 millimeters in length (2.93 inches); off the Canary Islands they were 60 millimeters (2.36 inches) and near the Azores between 35 millimeters (1.38 inches) and 45 millimeters (1.77 inches). Also, the larvae of the American freshwater eels (*Anguilla rostrata*) diminished from the eastern coast to the Bermudas.

The conclusion Schmidt drew was that the breeding grounds lay between the Azores and the Bermudas. Perhaps in the Sargasso Sea?

Then one day in 1922 Schmidt's ship the *Dana* was sailing southeast of Bermuda. There, at about 60° west longitude and 25° north latitude, the fine mesh nets dredged up thousands of tiny *Leptocephalus* larvae which were only about .4 of an inch long. They were evidently newly hatched.

Over the next four years Schmidt continued his work. Since then other marine biologists have filled in more of the gaps to complete the story, which is this:

Both the Atlantic and North American eels breed in the Sargasso Sea, between 55° and 68° west longitude. The Sargasso Sea is 20,000 feet deep, but the transparent pea-size eggs aren't laid on the bottom but at a depth of about 1500 feet where the temperature ranges from about 61°F. to 63°F. Here the water is warmer and saltier than anywhere else in the Atlantic. Buoyed by tiny droplets of oil, the incubating eggs float to the surface.

Once hatched, the larvae start on their long and separate journeys. The European larvae take three years to travel the 6000 miles to reach the mouths of freshwater rivers; the North American larvae take about a year. (The major difference in the two species is that the European eel has about 115 vertebrae, and its American cousin about 107. Some marine biologists claim that they both belong to the one species, *Anguilla anguilla*.)

Both use the warm Gulf Stream to make their way, and when they reach shore they metamorphose swiftly from a

three-inch larvae into elvers. At this stage of their development they are eel-shaped—narrow and rounded—and are a
bit shorter than their willow-leafed form because it shrinks
during the change. In Europe elvers are often called "glass
eels"—aptly enough because they are fairly transparent and
shiny.

The elvers migrate up the rivers, sometimes venturing
hundreds of miles from the parental sea. Males rarely swim
much above salt or brackish water; the eels you see beyond
that point are usually females. They are the ones that swim up
the rivers and make their way up tributaries and, on wet
nights, possibly slither overland to ponds and pools.

The eels, both males and females, don't become the
yellow eels we know until they reach sexual maturity and the
time for them to return once more to the sea. For European
eels this is usually when the females are twelve years old and
between four feet and five feet long; for males when they are
four to eight years old and about eighteen inches long. When
they start their journey they have put on a lot of fat; they'll
need it to sustain them because they do not eat on their long
trip. Like the Pacific Salmon and lampreys, their gut degener-
ates. They are much prized as food, particularly in Europe.
Many are netted and trapped when they're barely underway,
as they wriggle down the streams and rivers. At this time, too,
they change from yellow to silver, their sex glands begin to
develop, and their eyes grow larger. We can assume that their
great journey for about three years is at a vast depth—so deep
that we do not know the precise route they take.

How do they find their way? Perhaps by some sense
which measures the temperature and salinity of the water. Not
only are the waters of the Sargasso Sea the saltiest in the
Atlantic but the only deep water with a constant temperature
around 62°F.

It's not too hard to think of silver eels playing a kind of

blindman's buff, sensing variations in temperature and saltiness until the spawning grounds are found. There they breed and shortly afterwards die.

Why do the European eels make a 6000-mile journey when they could make a much shorter trip to North America? Marine biologists have often asked this question, and some have offered possible explanations. One is that continental drift may have played a part, that in the past the spawning ground was equidistant from both Europe and North America. Another possible explanation is that in the course of geological evolution the amount of deep water in the Atlantic with a suitable temperature decreased and also moved to the west, so that the European eels had to travel further and further in order to breed.

Besides the Sargasso Sea there are three other spawning grounds for freshwater eels. One is in the western North Pacific from which currents carry larvae to the coasts of China and Japan. Another is in the western South Pacific, near New Caledonia; currents carry eels to southeastern Australia and New Zealand.

A fourth spawning ground is in the Indian Ocean; currents carry eels to the north coast of Africa and to the western coast of Australia. These eels are also members of the family Anguillidae. (In all there are about 16 species of freshwater eels.)

Thus, freshwater eels, as we have seen, have four spawning grounds. No other exists, and only those areas whose shores are washed by currents from these areas have eels. For this reason there are no eels in the western regions of Canada and the United States.

Much has yet to be unraveled about the freshwater eels and their breeding. It's one of the great wonders in the story of eggs—of fish traveling thousands of miles to lay their eggs and reproduce.

16

WOOING WITH SOUND

BEFORE FISH CAN MATE they must find each other. Male salmon presumably find their home waters and their mates by smell; male sticklebacks, Siamese fighting fishes and others court mates with spectacular dances. New research dating from the end of World War II shows that sound may be the major sense in some marine courtships, and that some male fishes use sound much as male songbirds do—to deter rivals and attract females.

A dramatic happening in World War II made many zoologists revise some of our long-held ideas of the "silent" depths. (Perhaps we thought fishes were silent because they have no vocal chords?) In 1942 listeners on some of the U.S. Navy's ships were getting used to identifying the sounds of vessels picked up on their newly installed hydrophones. Suddenly they were deafened by a sound later described as resembling pneumatic drills attacking concrete. An alert was sounded, but it wasn't the Japanese or German Navy; the din had come from a large shoal of common croakers (*Micropogon undulatus*), members of the family of drumfishes (Sciaenidae) entering Chesapeake Bay on their spring spawning migration.

By the time the U.S. defense forces had located the cause of the racket, some 300 to 400 million croakers had entered the bay. The noise was on a principal frequency of 600 cycles per second during early June and was produced by young fish. In early July the principal frequency had dropped to 250 cycles per second, somewhere near middle *C* (256). The sounds were associated with feeding and produced mainly by soundmaking muscles operating on the swim bladders which served as resonators. Both sexes made the sounds, and like birds, there were also choruses at dawn and dusk.

But in other members of the drumfish family, (Sciaenidae) only the males have sound-producing muscles. This suggests that these sounds play their part in courtship and that, again like birds, these species of drumfishes use sound to defend territory.

The ancient Greeks and Romans knew all about these noisy sea basses, the Sciaenidae. The Greeks called them "grumblers," and the Romans "little crows."

Just as noisy is another family of fishes, the Pomadasidae, more commonly called grunts, and found in tropical waters throughout the world. Many are beautifully colored. They make sharp noises by grinding their throat teeth together.

Since the ear-shattering experience in Chesapeake Bay, biologists have found an ear-dazzling wealth of noises in the "silent" depths—burbling, buzzing and cackling; croaking, drumming and groaning; moaning, rumbling and squealing; quavering, smacking, whining, whispering and whistling. And they have discovered more fishes that use sound to deter rivals and attract females.

Fish, as biologists have discovered, have many ways of making sounds and have excellent hearing organs. (Water is, of course, an excellent conductor of sound—almost five times better than still air. Air transmits sound at about 1100 feet per second, depending on the temperature; sound in warm water travels at about 4800 feet per second.) When you throw out a baited hook and sinker you may be able to hear it at 100 feet, but a fish will probably hear it at 500 feet, assuming, of course that its hearing is the same as ours.

Not long ago, when trolling for rainbow trout on New Zealand lakes (I caught several six-pounders), I found it advisable to let out my fly on as much as 150 yards of line. This gave the trout a chance to recover from the shock of the throttled-back sound of the launch's motor. We weaved the launch from side to side so that the fly, carried deep down on the lead-weighted lines, would cover as wide an area as possible. And several times I got a strike by waiting for some minutes before slowly winding in my line with the motor stopped. Such is the sound-carrying ability of water that a six-pound depth charge exploded off Dakar in 1940 was recorded on sensitive hydrophones 3100 miles away in the Bahamas.

Many bony fishes have a direct link from the inner ear to the swim bladder. This, in effect, works much like a hydrophone. In addition, bony fishes have lateral line organs connected to the inner ear. These organs and the sensitive lateral line sense organs are responsive to water disturbances caused by other fishes. The lateral line sense organs are present in pores or canals and form a line along each side of the body. Others are arranged in lines on the head. Aquatic amphibians also have lateral line sense organs.

As well as making sounds with muscles connected to resonating swim bladders and grinding their teeth, fish make noises in other ways. Some do it by *stridulation* (rubbing one part of the skeleton against another). Triggerfishes (Balistidae) do so. So do filefishes (Monocanthus spp.), surgeonfishes (Acanthuroidea), sticklebacks and some catfishes (Siluridae).

Sea horses produce clicking noises when courting. They do so by lifting their heads suddenly. At the back of the head is a star-shaped bony crest (the coronet) which engages the rear edge of the skull. You will recall the earlier description of the courtship dances of the sea horses. During these they produce these snapping noises, first one and then the other fish, in a kind of underwater duet. When they finally embrace, the sounds grow louder and are nearly continuous.

Sounds produced by fishes range from about 50 cycles to 10,000 cycles per second and are usually emitted in bursts. Their function isn't fully understood. Some sounds play a part in courtship—the subject of this chapter. Other sounds probably help a school of fish to keep together. Thus, a school of anchovies (*Anchoviella choerostoma*) made no sounds when idling but uttered them when moving or changing direction.* These sounds could have been made by the motion of swimming. Sounds may help some catfishes swimming in murky water to communicate; one South American catfish makes a loud grunt you can hear a hundred feet away. In similar conditions a large family of African freshwater fishes, the mormyrids, use a kind of sonar; these fishes, some with elephantlike noses, emit an electrical field. But there is no evidence that electric signals play a conspicuous part in courtship, although they may well do so.

'   Some of the most stentorian sounds are produced by toadfishes, which have a heart-shaped swim bladder with a thick band of red muscle along two sides. The sounds made with this "loudspeaker" can be heard clearly out of water. If you are underwater and are close by, they can be almost deafening. "Measurements have shown that at a distance of two feet the sound output of a single toadfish may reach an intensity of 100 decibels, a value comparable to the noise of a

---

* N. B. Marshall in *The Life of Fishes* (New York: Universe Books, 1966).

riveting machine or a subway train." * One species, the oyster toadfish (*Opsanus tau*), which lives in shallow water along much of the Atlantic coast, is something of a virtuoso. It emits a foghorn sound at 140 cycles per second (almost an octave below middle *C*), with harmonics at 140 cycles-per-second intervals up to nearly 2000 cycles. The breeding season is in June and July. The females lay eggs in debris, even in tin cans or discarded shoes. Males guard the eggs until they hatch— from 10 to 26 days or so.

Toadfishes live on the sea floor and defend territories with a salvo of sounds. Noisy defense of territory is at its most vocal when a male is defending his nest with its eggs. He utters foghornlike cries and makes threatening gestures whenever another fish approaches.

Sounds loud enough to be heard out of the water are also made by squirrelfishes (Holocentridae) during spawning.

Another noisy fish is that close relative of the toadfishes, the famous singing midshipman, *Porichthys* spp., of the American Atlantic coast. They, too, have drumming swim bladders and are audible, even above the water, at some distance. There are accounts of loud choruses when many males gather together in the breeding season in the one area to find females. A male can produce a single boat-whistle blast—hence the popular name. Sound and threatening displays are used in defense of territories. Midshipmen have from 600 to 840 small organs under their bodies which produce light of such intensity that a zoologist found he could read a newspaper by it. Presumably the fishes use the lights to identify each other, and they may play some part in courtship.

All three senses, sound, smell and sight, play a part in the courtship of an eastern Atlantic goby, *Bathygobius soporator*.

---

* Dr. W. N. Tavolga, "Foghorn Sounds Beneath the Sea," *Natural History* 69 (1960): 44.

Males, who only "talk" when courting, bow their heads sharply and emit (no one knows how) low grunts. The noises could be to deter other males, because experiments show that when recorded sounds are played to the gobies only the males take any notice; the females are only responsive when they can see a male or males as well as hear the sounds.

Even the deepest parts of the sea aren't silent; sounds from 2000 fathoms have been recorded. Here, where no light from the sun penetrates and many fishes manufacture their own light, sound is probably increasingly important. (Little or no sunlight reaches below about 2500 feet.)

Significantly, about 30 species of rattail fishes (Macruridae) are able to produce sounds with strong drumming muscles on their swim bladders. Only the males can do so, presumably in order to find spawning females. What is interesting is that these fishes with long, tapering, scaly tails (which gives them their popular name) have well-developed inner ears. These doubtless serve the fishes well in the Stygian depths, but little is known about the volume of sound produced and heard by these fishes which are probably the most abundant inhabitants of the deep.

Besides using sound for courtship, some deep-sea fishes are thought to use echolocation by high-pitched sounds to maintain their level in the ocean, as well as to locate food and avoid enemies and obstacles. The U.S. oceanographic research ship *Atlantis*, when about 200 miles north of Puerto Rico, located on the hydrophones an unknown animal at a depth of 12,000 feet. The creature was emitting chirps lasting between two- and three-tenths of a second and pitched an octave above middle *C*. Regularly after each call an echo came clearly from the bottom, nearly 17,000 feet below.

Deep-sea fishes may possibly use luminous lights for courtship. Below 20,500 feet, marine creatures light the darkness almost continuously, much like glowworms, and flashes of light have been photographed at 12,000 feet. About two-thirds of the fishes of the depths and middle depths have light organs (photophores). Some of these fishes make their own light, which is remarkable enough; others, such as the rattail family, don't produce their own light but, incredibly, use that of luminous bacteria. The luminous bacteria are in the light gland and are, presumably, activated by the flow of blood which carries oxygen to the bacteria. Recorded by our eyes (we don't know how the fishes see it), the light flashes range from orange to yellow and to yellow-green and blue-green, a range that takes in almost half of the color wheel.

Many of the rattail fishes that swim near the ocean floor have a light organ on their bellies. Also, like other deep-sea fishes, they have large, highly sensitive eyes. Mr. N. B. Marshall, the English ichthyologist, thinks the lights may well play a leading part in courtship. Earlier we referred to the loud drumming these fishes make with their swim bladders and which travels a considerable distance. Just as male frogs do, the male rattails could lure the females to them. When the females are within 30 feet or so (the presumed range of the lights), then a luminous appeal could be linked to the sound courtship. "Once a male has met a mate, he may, like a male codfish, lead her away from the ocean floor to a level suitable for spawning," says Mr. Marshall in *The Life of Fishes.* "If so, the glowing of his ventral light could stimulate her to follow."

Colored lights may be used in wooing by male black dragonfishes of the genus *Idiacanthus* which have much larger ocular (cheek) lights than the females. The larvae are remarkable because they carry their comparatively large eyes on the tips of slender stalks; a tiny fish about two-thirds of an inch long has eyestalks about a third of its length. As the larvae grow, the stalks are slowly absorbed and are completely gone by the time full growth of one and a half inches is reached.

That *colored* lights may play a part in courtship is, of course, a speculation that only deep-sea television can resolve. Certainly in the sea there could be nothing more remarkable than a courtship with colored light.

Smell, too, could play a part in some courtships—particularly among fishes living in dark, stagnant waters. Little work has been done in this field. Some current research has shown that some male anabantids (the "labyrinth" fishes of Chapter 13) release into the water a substance called a pheromone that attracts females. The pheromone worked even when the female was unable to see the male in an adjoining glass tank. By using pheromones, zoologists have induced two different species to mate and produce hybrids—something they would be most unlikely to do in nature.

Ichthyologists suspect that pheromones could play a major part in the courtship of other fishes that inhabit dark, still waters—habitats where sight is at a discount and smell would help mating fishes to find each other.

It is only too true to say that love will find a way, whether it be with sight—or as this chapter has shown—with sound or smell.

17

PORTABLE BRIDEGROOMS

THROUGHOUT THIS BOOK we have seen fish in the spawning season finding each other and mating. But the male and female fishes in this chapter never need to seek each other.

These unique fishes which are always together in a state of permanent marriage belong to the suborder of Ceratioidea deep-sea anglerfishes. A female up to two or three feet long may have three or four tiny parasitic males attached to her. These "portable bridegrooms," about four inches long, are completely parasitic, fed by her bloodstream. Earlier they had been free-swimming. Then, on finding the female, they had clamped onto her body with their mouths. Her tissues broke down at the point of attachment, and each male literally became "one flesh." In a short time the internal organs of the males, except for the reproductive ones, degenerated. The males thus became completely dependent on the female's blood system for the rest of their lives.

Anglerfishes of the order Pediculati are famous for their

method of trapping their food with "fishing lines" on their snouts. In some the lure resembles a wriggling worm. Small fishes and prawns that are attracted are engulfed in a flash. In the Stygian gloom of the deep sea, the females fish with luminous lures.

During the second half of the nineteenth century, marine biologists were puzzled when their nets brought up from the depths only females of four families of deep-sea anglerfishes. They asked themselves were these creatures hermaphrodites? But dissection of the fishes showed no evidence that this was so.

About the same time, biologists found that among one order of deep-sea fishes, classified as the Aceratiides, there were apparently no females. But as these fishes were only a few inches long when mature, no one thought of connecting these tiny fishes with the gigantic female anglers. (Some specimens two feet long had been captured.)

The truth began to force itself on biologists early in this century when female anglers were fished up with three and four parasitic Aceratiides attached. At first, biologists offered the not unreasonable explanation that the Aceratiides were parasitic—after all, parasitism is not uncommon in the sea. But, increasingly, the facts mounted until they could not be ignored. More and more female anglers were dredged up with parasites from only Aceratiides. And this group wasn't parasitic on any other fishes. Moreover, the apparent parasites were frequently attached close to the female's urogenital vent. The logical inference was that females of four families of anglerfishes of the ocean's abysses had "portable bridegrooms" and that the Aceratiides had been wrongly classified and were not a separate order at all.

It makes very good sense for a female in the vast darkness to carry her mate or mates around with her. Finding each other in the gloomy depths must be a problem for other fishes.

Among these anglerfishes there is, of course, no courtship.

The males are, in effect, no more than portable, enormously developed sacs of milt, the discharge of which is triggered by the chemistry of the female's bloodstream.

Not much is known of the larval type of either males or females, or how the bridegrooms find their brides. It is just possible they find them by means of the luminous lures the females use for fishing the depths. These lights on their long rods may be death to small fishes, squids and prawns, but they could be love-lights for the tiny males. If this is so, then some of the ways of saying it with lights must be extraordinarily beautiful—and grotesque. One angler fish which Mr. William Beebe, the American marine biologist, caught in his nets at a depth of 3000 feet off Columbia, South America, had a purplish-blue light on the rod projecting from her nose and luminous bushy barbels hanging down her chin. Incidentally, her teeth were so long she couldn't close her mouth. Another had fleshy tubercles on its snout which produced intermittent white flashes. A stalk on top of its head bore other intermittent flashing lights and was topped with two thick, luminous tentacles. (Fishing "rods" in some species may be as long as the body.)

Perhaps it is a romantic notion that the males are lured by lights; they could be guided by their well-developed nasal organs. And they don't always home in on the bride where they'd be most effective. Biologists have dredged up female anglers with tiny husbands attached to the top of their heads or on the gill covers.

The seas have many strange and wonderful creatures but surely none more remarkable than the females who fish for their living with lights and carry their husbands around with them.

PART TWO

Invasion of the Land: Amphibians and Reptiles

18

FROGS: ARCHAIC VOICES

T HIS BOOK began with a look at the present-day amphibians and the way they reproduce themselves, mainly by eggs laid in water. Today's amphibians are descendants of the first vertebrate explorers of the land.

There are three orders: the Gymnophiona (caecilians which are legless, burrowing amphibians confined mainly to the warmer regions); the Caudata (salamanders); and the Salientia (frogs, toads and their close relatives).

In these orders are about 3000 species. Their common characteristics are:

- A wet, slimy skin;
- All are poikilothermic—that is, "cold-blooded" but with varying body temperatures according to their surroundings;
- Most lay their eggs in water;
- The adult is usually a land-dweller and breathes by means of both lungs and skin. (Some amphibians haven't lungs);
- In most the larva (called a tadpole) lives in the water and breathes by means of gills.

FIG. 6. *Some of the species described in Chapters 18 and 19, in approximately proportionate sizes. Left to right* (top): *edible frog of Europe; Zetek's frog of Panama, one of the arrow-poison frogs; female Surinam toad with eggs embedded in her back;* (bottom): (on leaf) *two color phases of greenhouse frog, Florida and West Indies; Darwin's frog of South America; male midwife toad of Europe carrying strings of eggs; bullfrog of North America.*

Amphibians have only been partly successful in the invasion of the land, although in developing a three-chambered heart they have advanced on fishes with a two-chambered heart. They lose water through their skins and, therefore, cannot live in dry areas. And most must return to freshwater in order to breed. Amphibians have not adapted to salt water. As one expert puts it, the amphibian is little more than a peculiar type of fish which is capable of walking on land. As it were, they have merely established bridgeheads on the land since that time long ago in the Devonian period when the first amphibians developed. The conquest of the land could only be achieved by the reptiles which found a way of reproduction by a special kind of egg that freed them from dependence on water to breed.

The amphibians had their moment of evolutionary glory in the Carboniferous period which followed the Devonian when they were the dominant land animals. One amphibian called *Eryops* was six feet long, was a fish-eater, and looked rather like a large clumsy frog. Others were eight feet long with tails—far bigger than their descendants living today; others were small creatures. Their story is mainly lost because the fossil record is scanty. We shall take a close look at the tailless amphibians (frogs and toads) and their eggs. I've chosen them because we are more familiar with frogs and toads than with the other amphibians, even though they are not as ancient in their descent as other amphibians such as the little-known wormlike caecilians and some of the salamanders.

When I looked into my ornamental pool in the morning I found what the night's din had led me to expect—masses of frogs' eggs on the bottom among the stalks of the water lilies. Each tiny life-germ was in its protective capsule of albuminous jelly and attached to the nutritive store of yolk on which the tadpole was to grow. The egg's outer covering of jelly was taking in water, and before long the masses of eggs, each globule of jelly swollen to about the size of a pea, would float to the surface. The frog spawn, with which you are probably familiar, would cover most of the pond's surface and look much like a frothy jelly with tiny black spots in it. The black spots are, of course, the germ spot or true egg.

My pond was a new one, but it hadn't taken the frogs long to find it. "Wank-wank-wank!" had gone the chorus of some males. And "ping-pong-ping-pong!" And "bonk-bonk-bonk!" There were several species there; only an expert or a female frog could have sorted them. Most were tree frogs, but the banjolike "bonk-bonk-bonk!" was the love song of bullfrogs.

Monotonous, insistent, inharmonious, the sounds that had kept me from falling off to sleep were also, I told myself, some of the most romantic and wonderful in the world of nature. Sounds much like these had probably been the first sounds made by the voices of living creatures in a hitherto largely silent world, when males of the earliest froglike amphibians began to gather together for mating and sought to lure the females to them with croaks and chirps. Before the first amphibians had emerged, there were, of course, inanimate noises—of wind, of waves crashing on beaches, the roar of thunder, the rumble of volcanoes—but the only sounds produced by living creatures were the nonvocal ones of insects.

So, however harsh and discordant, the night noises of frogs are also archaic and evocative of mystery. Sounds much like these filled the land long before birds sang. And, although you'll find no mention of it in the more orthodox history books, the uncouth serenades of male frogs probably helped to bring about the bloody holocaust of the French Revolution. On the great estates of the nobility during spring and summer the peasants had to beat the streams and ponds all night through so their lords and ladies could sleep! Lack of sleep undoubtedly made the French peasants irritable and could have been one of the factors that precipitated their revolt.

In the previous chapter we saw how the first amphibians left the waters and came ashore, exchanging one element for another. They performed the slow miracle of ceasing to take oxygen from the water and acquired the means of absorbing it from the air. From the first amphibians evolved the reptiles, the birds and the mammals, including ourselves.

The miracle—rather speeded up—of switching from gills to lungs is still performed today by the 2600 species of frogs and toads and their near relatives during the development from tadpole to adult. All make up the order Salientia of tailless amphibians. ("Salientia" means "leaping.")

The eggs in my pond underwent several rapid changes. The first was when the round, black egg developed a "waist" and looked like a tiny dumbbell. (In fact, it's called the dumbbell stage.) Then it came to resemble a comma, and then, after about 14 days, the tadpole freed itself from the jelly, looking much more like a fish than a future land dweller. It anchored itself to the jelly with a sucker under its head. It couldn't eat as yet—it had no mouth—but lived on the rich food of the yolk. After two or three days most of the yolk had been absorbed, and the tiny tadpole had grown external gills, and a mouth with horny jaws and lips. It is a curious mouth, rather like that of a lamprey. With it the hungry tadpole feeds on algae which it rasps away from the plants on which it grows.

The changes that follow—and which you have probably watched—are startling. Organs grow inside the body; the outer gills disappear, to be replaced by two others inside the body.

After about four weeks the buds of two rear limbs show and grow slowly. As yet, they are not used; the tadpole moves as a fish does with lateral thrusts of its rudderlike tail.

As it continues to develop, the tadpole stops feeding on algae and eats animal food. (You may have seen tadpoles feeding on a dead fish, which they'll strip clean. If you keep them in an aquarium, they like to feed on the remnants of meat on a bone.)

The days and weeks pass—the metamorphosis of a tadpole to frog depends greatly on the temperature. Seven to nine weeks after hatching, great changes have taken place. Lungs have begun to form to replace the gills which become increasingly less able to supply the tadpole with sufficient oxygen; it has to swim continually to the surface to gulp in air. The lungs form, and for a time the tadpole can breathe in both water and air. The front legs bud, grow and thrust themselves out through what were the openings that once served the internal gills. Unlike the toes of the hind limbs, these forefeet are not webbed. The tail, meanwhile, has been growing smaller. It does not drop off, as many believe, but is absorbed. Our tadpole now is on the verge of the greatest change of all. It ceases to eat so that its mouth, tongue and jaws can develop. Teeth grow on the palate as well as on the edges of the top jaw. Don't worry, the creature won't starve while this is happening; it is living on the rich store of food in its now stubby, fat tail, which daily gets smaller.

Then one day—about ten to twelve weeks from the time of hatching—the tadpole's skin splits and a little tailed frog emerges. The froglet is about the size of a button, ready to feed on tiny insects, and perfect except for that knob of a tail. That, too, will disappear in a few days, completely absorbed into the body. But while it remains it can be a deathtrap for its tiny owner when he leaves the water for the first time. The tail is moist and can stick to dried surfaces, tethering its owner until he dies, sucked of his life-giving water by the sun. In addition to their lungs, frogs, like other amphibians, use their moist skins for breathing. Water is lost through his skin, and frogs can dry out rapidly.

Thus, the sun is one great enemy of frogs, who must seek moist places. They have many others—fish, birds, snakes. For this reason, frogs are prodigal with their eggs; a female will lay on average about 5000 eggs during each breeding season. If even a tenth hatched and grew to adult frogs, there'd be a bigger plague than ever harassed Pharaoh in Egypt. Some idea of the high mortality of frogs and toads may be gained by the figures recorded of the giant South American marine toad (*Bufo marinus,* six to eight inches long) which sugar planters all over the world have imported to check beetles. In Queensland, Australia, 37 females in experimental ponds laid over 1.5 million eggs during the breeding season, but only 62,000 toadlets were netted for release in the canefields. On average each female had laid 45,000 eggs.

The metamorphosis from tadpole to frog that we've looked at is typical of most of the 2600 species of Salienta. There are, however, some remarkable exceptions, such as froglets which never pass through a tadpole stage, and some which are born alive, which we'll look at later.

As Doris M. Cochran, Curator of Reptiles and Amphibians, United States National Museum, Smithsonian Institution, remarks in her book *Living Amphibians of the World*, our language is singularly poor in common names for the 18 families of tailless amphibians. "Frogs" and "toads" are the only English names in popular use for all the different species of these creatures. In strict terms, the name "frog" is only correctly used for members of the genus *Rana*, and "toad" is only properly used when it is applied to the genus *Bufo* and near relatives. But in common usage we apply "frog" and "toad" to many creatures that are not related to either the genus *Rana* or the genus *Bufo*.

What biologists call the true frogs—those of the genus *Rana*—are found on all continents: all over North America; in Central America and the northern part of Central America; in all of Europe and Asia; in Africa except the very cold or very dry regions; and in the northern parts of Australia. None exist in New Zealand.

Typical true frogs include the common American bullfrog, *Rana catesbeiana,* found east of the Rockies in the United States. (See Figure 6.) Other true frogs are the common frog of the temperate regions of Europe and Asia, *Rana temporaria,* and the edible frog (See Figure 6) of Europe, *R. esculenta.* (From Roman times, gourmets have eaten the legs of the edible frog. They taste rather like chicken and have probably always been a luxury—because it takes a lot of legs of these four-inch green frogs to provide a meal. The edible frog was introduced into the United Kingdom well over 100 years ago.)

All these frogs have some features in common, such as

slender, streamlined bodies, pointed heads, protruding eyes, webbed toes on the hind feet and powerful leg muscles that enable them to leap away from enemies. All have small teeth in the upper jaw, on the palate as well as on the edge. All, too, have long tongues with which they capture their insect prey. These forked tongues are sticky on the ends and are one of the most efficient food-gathering weapons in nature. A true frog may have a tongue nearly as long as its body. It is rooted at the front of the mouth and near the bottom lip to gain extra length, and the tip of the tongue hangs down its owner's throat. Moreover, the tongue is so elastic that when flicked at an insect it can stretch to about twice its resting length. In action it is lightning fast. Clocked by a high-speed camera, a bullfrog's tongue emerged in 0.05 seconds and whipped around its insect prey much as a weighted rope would spin round a post.

Many frogs jump for their prey. In tests, bullfrogs jumped at moving prey ten inches away and four to five inches above the water. From leap to splashdown (with the gulped-down insect) took only 0.3 seconds! Frogs and toads will only strike at live and moving prey. They are, of course, invaluable in controlling insects. Tree frogs (family Hylidae) will leap onto slender leaves after insects, but though they have disc-shaped pads on their toes to grip the smooth surfaces of leaves, they sometimes fall when doing so. Yet, even if they should fall to the forest floor, they are unlikely to be hurt. And while falling, they are often able to navigate themselves towards a nearby branch. Experiments show that even the common American bullfrog in a swift jump after an insect can change direction in midair by the use of its webbed hindfeet which are used as rudders. (Bullfrogs have also been known to vary their diet of insects with small birds and other frogs and even to swallow small snakes longer than themselves.)

We can recognize most toads fairly easily by their fat bodies, short legs and hopping gait. They have drier, thicker skins than frogs. They are not as athletic as frogs and rely, not on speed, but on the deterrent effects of an irritating poison which they emit from parotoid glands behind the eyes and from smaller glands all over their bodies. (Most frogs don't have these glands.) Any animal who has once seized a toad in its jaws is unlikely to repeat its mistake—the poison burns the mouth acutely. Some large toads carry enough poison to kill a large dog. Doris M. Cochran records that a police dog died after seizing in its mouth a large Colorado River toad, *Bufo alvarius*.

The poisonous qualities of toads have been known for a long time. You may remember what the banished duke says in Shakespeare's *As You Like It*:

> Sweet are the uses of adversity,
> Which, like the toad, ugly and venomous,
> Wears yet a precious jewel in his head.

Fortunately for toads, they don't carry precious jewels. However, the poison the toad secretes could turn out to be a "precious jewel" in disguise. Analysis of these poisons reveals at least three whose action is similar to that of digitalis, which doctors use to treat some kinds of heart trouble. (For centuries Chinese physicians have used toad skin in their medicine for heart ailments.) Thus, these poisons could be of value in modern medicine. The toads' poisons, acting much as digitalis does but to a greater and hence harmful degree, increase the muscular tone of the heart to the point of stopping it, weaken breathing and bring on muscular paralysis.

To divert briefly from toads and their poisons, the most spectacular poisoners are not toads but the arrow-poison frogs of the Central and South American forests. (See Zetek's frog, Figure 6.) For centuries the Indians have used the poisons of small, brightly colored frogs belonging to the family Dendro- batidae to poison the tips of their arrows. The Indians gather this poison, secreted from the frogs' skins, in jars into which they dip the tips of their arrows, after which the tips are allowed to dry. Carrying a concentrated dose, the poisoned tips will paralyze a monkey or bird within a few seconds.

The poison serves its frog owners well. Many are quite fearless because they've learned that no enemy will try to eat them.

All amphibians have at least a trace of poison in their skins. In most frogs there isn't much—they don't need it and rely on speed of leaping to get themselves out of danger. Only the fat, clumsy toads need poison as a universal deterrent.

Other differences between toads and frogs are that toads have drier, warty skins and, unlike frogs, no teeth in either jaw. (But it's an old wives' tale that toads give you warts.) Also, most toads lay their eggs in strings instead of in an amorphous mass of jelly.

In the opening of this chapter we watched the develop- ment from egg through tadpole to tiny froglet. If the tiny froglet survives its numerous enemies, it will reach sexual maturity and mate. In some of the true frogs sexual maturity may be reached in 18 months; the American bullfrog may take up to five years.

Courtship among frogs is a noisy affair—to our ears anyway; female frogs, presumably, like the sounds the males make and are lured to them. They breed at night because, as Professor N. Z. Berrill points out, in *Sex and the Nature of Things,* they "are among the meek and humble of this earth." They're safer at night—from predators and from their great enemy the hot sun. In their mating journeys frogs and toads may travel up to a mile, although some don't move far from the water which gave them life; the American bullfrog, for instance, spends its life within 100 to 300 yards of its home pond. Thus forced to woo by night, it is hard to find one another using their eyes, so the male frogs' serenades of croaks and peeps serve as sound beacons, as it were, on which female frogs can set their course.

Although unseen by most of us, the nocturnal spring migrations of frogs (and toads) are as remarkable as those of birds. Male frogs (and toads) emerge first from their winter hibernation and make their way in most instances to the very same ponds in which they were bred. Usually they travel along streams—when nights are damp they may go overland. How do they find their way to their home waters? Very probably by smell, according to the English expert Dr. Maxwell Savage.* He thinks that the smell of minute plants in the home water is carried on the air to the advancing amphibians, and that the particular smells are recognized by the frogs.

In England the famous naturalist Maxwell Knight has observed the extraordinary spectacle of an army of migrating common toads (*Bufo bufo*) on a march of at least a mile, from their hibernation sanctuaries to their birthplace, and clambering over every obstacle in their way. In *Frogs, Toads and Newts in Britain*, he writes:

> There is a place in Kent where I have watched migratory toads over a period of years. The lake in which they spawn lies in the grounds of a castle, and between the lake and the road along which each year the toads journey, there is an old, ivy- and creeper-covered wall which, at certain places, is about four feet or so in height. There is no way into the castle grounds along this road other than by climbing the wall. I have seen the

---

* Savage is quoted indirectly by Maxwell Knight in *Frogs, Toads and Newts in Britain* (Leicester: Brockhampton Press).

toads—some of them already with their mates on their backs—climbing up this wall just as you would climb up a taller wall with stronger creepers. The toads use their front and back legs, their toes seeking to grasp some stem or crack in the wall to enable them to reach the top. When they get there, they just flop over the edge and continue for about twenty yards until they reach the lake.

Perhaps you've seen a bullfrog croaking, with his vocal sac puffed up like a balloon? The purpose of this sac is not understood by many people, including naturalists. "The male frog calls by blowing out his dewlap and letting it go with a burp and depending on the size, you hear the peep of a peeper or the grunt of a bull," says one naturalist rather picturesquely.

It is not like that at all. The vocal sac or sacs—some frogs have two—are sounding boards to give resonance to the calls the frog produces. Air driven back and forth between lungs and closed mouth makes his vocal cords vibrate, and the sac amplifies the sounds. Note the closed mouth—a frog can call underwater, and some species do. Females have vocal chords but are generally silent. Females of some species sometimes use a chirplike call to tell a male that they are not ready for mating.

The croak of a male frog serves much the same purpose as the song of the male songbird during spring; it informs females that he is ready to mate with anyone willing, and it also tells other males that a particular area is occupied.

If another male, croaking a challenge, moves too far into the area and is not checked by the calls of the occupier of the site, then the resident answers with a loud and distinctive rivalry call. In some experiments in Australia a Melbourne zoologist, Dr. Murray Littlejohn, recorded the mating calls of an Australian bullfrog, *Limnodynastes dorsalis,* and played them back to him. When the recorded calls reached a critical point of loudness, the bullfrog answered with his rivalry calls. Dr.

Littlejohn noted* that the banjolike "bonks" of the bullfrogs were the mating calls, and the long growls were the rivalry calls.

Just as with birds, the rivalry call is often sufficient to check an intruder; but among some species fierce fights may occur among rival males. Male marsh frogs (*R. ridibunda*), of Europe, try to bite the vocal sacs of rivals.

A third type of call is produced when a male makes vocal contact with another male—or a female that isn't ready for breeding. The other male or the female answers with what zoologists call the "release call." It is a useful signal to the first male not to waste his time.

The fourth call is one you may have heard—a loud scream when a frog is taken by a predator. It's startlingly loud and often helps the victim—the predator may be frightened into releasing the frog. And it's the only call made with the mouth open.

Dr. Littlejohn also found distinct differences in the calls of various species—differences which in most instances we would not detect. Our knowledge of frog calls generally starts and finishes with the recognition of those which are very distinctive—such as, for example, the carpenter frog (*Rana virgatipes*) of North America, whose call sounds like two carpenters hammering nails, not quite in unison. Frog calls have been variously described as twanging, sonorous, bell-like, clicking, etc.

Among frogs, like calls to like. The distinctive call of a male of her species lures the female to the swamp or stream where he waits. The male grasps the female from behind and under her armpits by means of the thickened pads on his thumbs. These pads grow there only during the breeding season and are, doubtless, a great help in clinging to his

---

* In *Wildlife in Southeastern Australia*, an Australian Broadcasting Commission pamphlet.

slippery bride (who, by the way, is usually the larger of the two). When the male is clasping the female they are said to be in *amplexus*. The male's embrace stimulates her to extrude her eggs. The two may be together in amplexus for hours until she has laid her eggs underwater. He ejects sperm from his cloaca and fertilizes them as they descend. The fertilization of the eggs resembles that of the frog's ancestors, the fishes, because male frogs have no penis. They are ancient eggs, "almost as antique in kind as those of lampreys," as Professor Berrill puts it. Toads' eggs are equally ancient.

With the mating we have come full circle—you will remember we began with the mass of eggs in my ornamental pond which hatched into tadpoles and developed into froglets. That is the normal story. But within the vast company of frogs and toads there are exceptions to this pattern. Some of these are the subject of the next chapter.

19

FROGS THAT ARE DIFFERENT

SOME FROGS, such as those belonging to the genus *Eleutherodactylus,* skip the free-living tadpole stage —that is, the tadpole develops inside the egg and emerges as a perfect froglet. The froglet has a tiny egg tooth on the tip of its snout to help slit the egg membrane. One member of this genus is the greenhouse frog (*E. planirostris*) of the southeastern United States, so called because it frequently lives under pots in greenhouses. (See Figure 6, Chapter 18.) Encapsuled tadpoles do not have to run the gauntlet of water-dwelling enemies as other tadpoles do, so, not surprisingly, the mother frog lays far fewer eggs than most spawning frogs do—about 25 or even less.

As we saw earlier, female frogs and toads are generally prolific egg-layers—they have to be because mostly they take little care of the incubating eggs and emerging young. But there are some startling exceptions. There is, for instance, the hard-working Brazilian female tree frog, *Hyla faber* (the smith frog). She builds a little dam for her eggs in the shallow end of

147

a pool by diving to the bottom and emerging repeatedly with lumps of mud on her head. These she deposits into a circular wall, up to four inches high, thus forming a tiny reservoir. There she lays her eggs, after which she and her mate guard the nursery until the tiny tadpoles have changed into small frogs.

A female tree frog in the Amazon Valley steals bees' wax from stingless bees and uses it to make a watertight breeding pool in a hollow in a tree trunk, which she guards.

Other tropical tree frogs make breeding pools by drawing leaves together. Some Asian frogs lay their eggs in a ball of beaten foam attached to a tree overhanging a pool or in rice paddies. The foam is made from a secretion of the females and is whipped up by the hind legs of both parents during the egg-laying. The manufacture of the airy nursery may take several hours, during which time the male may squirt water over it to help things along. Over a period of some days the emerging tadpoles make their way to the bottom of the nursery and into the water in the pool or rice paddy to continue the metamorphosis into frogs.

All these tadpoles have better chances of becoming froglets than do other tadpoles. Those of a tiny inch-long frog found in Chile and Southern Argentina have an even safer start in life because they grow to froglet stage in the safety of the male's enormously enlarged vocal sacs. Charles Darwin discovered this remarkable mouthbreeding frog which is named after him, *Rhinoderma darwinii*. (See Figure 6, Chapter 18.) The males wait around a clutch of 20 to 30 eggs laid by a female, guarding them for 10 or 20 days until they are almost hatched. Then each of the males picks up several eggs at a time with his tongue and guides them into the slit of his vocal sac. In this way each male may take in from 5 to 15 eggs, which develop into tadpoles and ultimately froglets within the sanctuary of the vocal sac. The froglets that emerge—sometimes from their own father's vocal sac—are about half an inch long.

Another careful amphibian father is the midwife toad of Europe (*Alytes obstetricans*). I mentioned earlier that female toads lay their eggs in beadlike strings. With these toads, mating takes place on land, and the male midwife toad winds these strings of 20 to 60 eggs around the lower part of his body and keeps them moist by dipping them in water intermittently for about a month until the tadpoles are ready to emerge. While guarding the eggs he generally hides during the day in a cool spot and comes out in the evening to enter the water.

The female Surinam toad (*Pipa pipa*) has developed one of the safest methods of raising her young; she carries her nursery on her back. (See Figure 6, Chapter 18.) When she is breeding, the skin on her back is abnormally thick and soft. During mating, up to 60 eggs are maneuvered onto her back by the embracing male and are pressed by him into the spongy tissue. After fertilization the eggs develop from tadpole to froglet in tiny sealed cells on their mother's back. The female looks, as Doris M. Cochran puts it in *Living Amphibians of the World*, like a rectangular somewhat scorched pancake, covered with lobes and fringes. Incidentally, the Surinam toad belongs to one of the strangest of frog families, the Pipidae; they have no tongues and scavenge in the mud with their long, slender fingers.

A male nursery technique has been adopted by tree frogs in the Seychelles, a habitat that lacks stagnant water. The female lays her eggs under wet leaves, and the emerging tadpoles attach themselves to the male's back, adhering by their sucker mouths and the stickiness of their tails. There they develop into frogs.

There are other remarkable methods that frogs have acquired to ensure greater safety for their tadpoles. Rohde's treefrog, *Phyllomedusa rohdei*, of South America, tries to ensure greater safety for its eggs by gluing them to a broad leaf overhanging the water. When the eggs have developed to the early tadpole stage, they drop into the water.

A genus of tree frogs, *Gastrotheca*, are called "marsupial frogs" because females hatch their eggs in a pouch on their backs. In one species, the male places the eggs in the pouch. At tadpole stage, the female releases them in water. In another species, the female raises her hind legs so the fertilized eggs roll into the pouch. When the froglets are ready to emerge, the mother raises a hind foot to her back to tug apart the slitlike opening of the pouch and help the fragile young escape.

Maternal care is also exercised by a tiny frog, *Cophilaxus pansus*, that lives in the alpine grass of New Guinea at altitudes around 10,000 feet. The female lays, on land, a number of large eggs joined by a string of mucous. The eggs are guarded by the mother until they hatch as fully formed froglets. For a short time the froglets are carried about on their mother's back.

Perhaps the safest start in life is given to youngsters belonging to two species of the genus *Nectophrynoides* in Africa. Female toads of this unique genus bear their young alive. They are the only members of the order Salientia that do. The females are fertilized internally and may carry up to 100 larvae at a time.

One fascinating example of adaptation to arid conditions are some African tadpoles which, having adapted to changing into froglets in soil with little moisture, drown if you put them in water! The female Rattray frog (*Anhydrophryne rattrayi*) lays her 20 large eggs in a small hole in the ground. Helped by a little moisture in the soil, the tadpoles undergo the metamorphosis to froglets before emerging.

Gray's frog (*Rana grayi*), of South Africa, likewise lays its eggs in small holes in the ground. The tadpoles develop inside the hard egg capsule, absorbing moisture from the night dews and occasional light rain. Several weeks may pass in this fashion until the winter rains come and the well-advanced tadpoles wriggle free of the eggs.

There's yet another unusual frog whose breeding pattern is absurdly different, though it is doubtful if there's any advantage gained. Adult frogs and toads are usually much larger than they were as tadpoles, but it's the reverse with the paradoxical frog (*Pseudis paradoxa*) of tropical South America. The tadpole may grow to ten inches long, then shrink steadily into a tiny frog which, when mature, is about three inches long!

Frogs and toads are widely distributed, but more abundant in tropical and subtropical regions than elsewhere. They are rare in very cold or very hot and dry regions. In Sweden one species lives well within the Arctic Circle. In the arid regions of Africa and Australia some species live a protected life in the cool galleries of termite nests and prey on their hosts. Some frog species can even survive in hot springs where the temperature is 100° F. One genus of toads, *Breviceps*, in Africa, is so free of dependence on water that they have lost the ability to swim and their froglets develop entirely inside the egg.

In cold regions frogs and toads hibernate—sometimes in the mud of ponds. Hence the riddle sometimes posed: What is it which first lives in water and drowns in air, lives next in air and drowns in water, then buries itself at the bottom of water and breathes nothing?

Just as remarkable as hibernation is aestivation. This form of summer sleeping, not quite as deep as hibernation, enables some frogs to survive months and even years of drought. They seal themselves into moist little chambers in the drying mud of a watercourse or a pond and sleep until heavy rain calls them forth to utter their strident mating calls. Aestivating frogs in the arid inland of Australia store considerable amounts of water. A notable one is the water-holding frog (*Cyclorana platycephalus*), which stores water in its bladder and body cavity to tide it over months of drought. These "reservoir" frogs in the past have saved the lives of desert aborigines who dug them out of their sanctuaries in the mud of dried-up pools.

Some species of spadefoot toads in the more arid regions of the United States have adapted themselves for survival not merely by aestivating but by being ready to breed within a few hours after rain has fallen and filled the dried-up pools to a depth of two feet or so. Males gather in the pools and utter calls so loud that human ears can hear them for two miles. Some female spadefoots have been discovered to respond to the loud love calls from as far as 600 yards.

Spadefoot toads get their popular name from the crescent-shaped bony projection on each side of the front feet—the "spade" with which they burrow efficiently into the soil.

Hurter's spadefoot (*Scaphiopus hurteri*) sets something of a record in the swift raising of a family. The eggs hatch quickly, and the froglets are often ready for life ashore in 12 days! (The rate of growth depends on temperature and the amount of food in the pool.)

Various members of the spadefoot family (Pelobatidae) are widely distributed not only in North America but over Europe, northwestern Africa and southern Asia.

It's the emerging of aestivating frogs which gives rise to most of the stories about frogs raining from the sky. (It is just possible that whirlwinds could snatch up frogs from water at any time of the year and deposit them miles away.)

Another legend about frogs and toads is that they can live

for ages sealed up in coal and rocks. This is, of course, nonsense. First of all, frogs and toads are not exceptionally long-lived—perhaps 10 to 15 years for some of the longest-lived species. A naturalist who sealed up some frogs without food and without renewed supplies of air found that they lived for only a few months. However, frogs and toads are sometimes found in pockets and crevices in rocks and possibly have lived there for months or even years. The simple explanation is that they crawled there when tiny, lived on insects and eventually grew too big to escape from their rock prisons.

In summary, the Salientia, as we've seen, have made only a modest conquest of the land; most are captives of freshwater or damp areas and must return to the water or a damp place to breed. (No Salientia—or other amphibians—have adapted to salt water.) Their eggs for the most part are delicate; they lack a protective covering and dry out when exposed to the air; further, they usually have only a modest food supply (yolk), and the young must generally undergo a larval (tadpole) stage during which they breathe by gills.

Frogs, like the other amphibians we'll look at in the next chapter, are in constant danger of dehydration, lacking the tough protective skins of their descendants the reptiles. And the majority of frogs and toads exercise no parental care of their eggs.

The amphibians are, as Dr. Romer declares, a defeated group. They were the first of the vertebrates to emerge from the waters onto the land, but they were not destined to complete the conquest. They were at first abundant and dominated the land—or the fringes of it—in Carboniferous days. Some grew quite large, such as *Eryops*, which was six feet long and resembled a large-headed frog in general appearance. Others rather resembled crocodiles, and others, again, were small and snakelike. They were all to give way in turn to higher forms of life, leaving a few modest latter-day descendants, such as the frogs and toads and those we shall look at in the next chapter, the wormlike caecilians and the salamanders and newts.

Nonetheless, with the evolution of the amphibians the basic body plan of land-dwelling vertebrates had been established. This included the tetrapod (four-limbed) body of two pairs of pentadactyl limbs, crucial to locomotion on land. The basic body plan also includes a spine, limb girdles, and internal organs adapted to life on land.

20

CAECILIANS

THE FIRST ORDER (Gymnophiona) of living amphibians, the wormlike caecilians, are very much among the meek and humble of this world. Even if you live in those tropical and temperate regions where they're quite common, it's highly probable you've never seen one, and you probably never will unless heavy rains wash them from their burrows. Because of their retiring natures they are seldom shown in zoos or discussed in any except the most comprehensive textbooks—and not very much even there because we know little about their way of life.

When a caecilian (pronounced "see-sil-ian") is washed to the surface by heavy flooding, it's likely to be killed by someone who mistakes it for a snake, although they resemble a large earthworm, if anything. All caecilians are limbless.

The lineage of caecilians is very much more ancient than that of frogs, as is shown by the fact that they have no trace of a pelvic girdle. They also have other primitive characteristics: In addition to lacking eyelids, many species have small bony scales which are hidden in the slimy skin; each vertebrae is shaped like an hourglass, showing their descent from primitive amphibians and from fishlike ancestors.

They are all in a single family, the Caeciliidae, with 16 genera and 75 species; you'll find most of them in Central and South America (46 species), 23 in Africa and the Seychelles Islands, and 6 in Asia. The Australian continent has none.

Caecilians look very much like large-size earthworms. Their long cylindrical bodies are slimy and divided into folds or rings. One Colombian species grows to four and a half feet; others grow to three feet; smaller ones reach only seven inches. Most are of a dull black color with lighter bellies, but one is a lovely blue shade, another a gay yellow, and another a purplish black.

Most are probably carnivorous, feeding on earthworms and insect larvae. Adults are generally terrestrial (land-dwelling); there have been reports of some species found swimming.

Most lay eggs; in several species the female coils around the eggs until they hatch. Because they are amphibians, most young must undergo a larval stage and develop in the water. But some develop completely inside the egg and a few are born alive.

Not a great deal is known about their mating. Probably they find each other in their slimy burrows by touch and smell. They have eyes, but mostly they are small, degenerate eyes and useless because they are covered by skin or bone. Eyes, of course, aren't necessary if you live in dark burrows. Instead, a caecilian has a sense organ—a tentacle which it protrudes from its head or lowers back into a groove between its eyes. It is both an organ of touch and smell and is connected to the nasal passage and to the creature's Jacobson organ. This is a cavity richly supplied with nerve endings used for smell. The tentacle gathers minute traces from odor-producing objects and conveys them to the Jacobson's organ. (Snakes, too, have this remarkable organ; the snake's double-tipped tongue gathers scent particles from the air and ground and conveys them to two pits in the reptile's mouth.) The caecilian uses it to investigate an odor-producing object. The organ is named after its discoverer in 1809, the Danish anatomist Ludwig Levin Jacobson.

Caecilians, unlike other amphibians, achieve true copulation. The sides of the vent of the male can be thrust out and used for pairing with the female. Here is an instance of a first step toward a primitive "penis." (The development of a penis in male reptiles will be dealt with in a later chapter.) Caecilian eggs are thus fertilized internally.

Although of ancient lineage, caecilians are highly specialized animals—that is, they have undergone many changes in the course of evolution in order to adapt to their subterranean life. They are vastly different from their first ancestors. Attempts to trace their evolution have been handicapped by the absence of fossils that resemble them. The resulting picture can be compared to a jigsaw puzzle with a large gap of over 200 million years. The fossils that most resemble the living caecilians are of certain wormlike types of amphibians common in the Permian period about 230 million years ago. After that, there's no fossil record.

21

SALAMANDERS AND NEWTS

SALAMANDERS AND NEWTS are much better known than caecilians, although in their native habitats they are shy and secretive. They are favorite aquarium subjects, delighting us with their beautiful colors and courtship water dances. They are kept, too, in many zoological laboratories for studies of tissue transplantation, the working of pituitary and thyroid glands, the regeneration of limbs and other experiments.

Some 220 species and subspecies belonging to about 60 genera and 8 different families make up this third order of amphibians, the Caudata, or, as it is sometimes called, the Urodela. The order includes the salamanders and newts, lungless salamanders, axolotls, giant salamanders, amphiumas, olms, mud puppies and sirens. But for simplicity's sake we refer to them generally as salamanders and newts.

All members of the order Caudata are rather similar, with elongated bodies, short heads and long tails. The tails often have crests on the upper and lower edges. Their four limbs are usually about the same size; some lack hind limbs. The front legs have three to four toes—generally four—and the hind legs two to five toes—usually five. The limbs, however, are relatively feeble. When salamanders walk, their bodies are thrown into the sinuous curves with which their fish ancestors thrust themselves forward by pressure on the water.

Their skins are thin and without scales, thus their physical needs are much like those of frogs—that is, they need dampness and moisture for their porous skins and usually freshwater in which to lay their eggs. Also like frogs, they eat insects and worms. The skin of most amphibians is soft and thin—so much so that in some species the tips of the ribs frequently emerge through the skin, but presumably cause no ill effects. Since the skin does not provide sufficient protection against the desiccation caused by sun and wind, most salamanders and newts—again, like frogs—are nocturnal.

Salamanders and newts are sometimes mistaken for lizards, but their moist, scaleless skins readily distinguish them from lizards, which have dry, scaly skins.

Although their distribution is much the same as frogs, they are not nearly so successful. There are fewer of them. (Compare their 220 species as against 2600 species of frogs, toads and their relatives.) They live almost exclusively in the temperate regions of the Northern Hemisphere—about half of the known species inhabit North America; Australia and New Zealand, both Southern Hemisphere countries, have none.

Most salamanders and newts are terrestrial and live under the forest debris; a few are tree dwellers; some others are aquatic. A few remarkable ones live in caves in total darkness and have become blind or nearly blind.

Salamanders and newts lay two kinds of eggs. The first is the presumably more primitive pigmented type with a small yolk, and the second is the presumably more advanced unpigmented type with a large yolk. The salamanders who lay the first kind abandon them to their fate much as many fishes do, making no attempt to hide them. Those who lay the apparently more developed eggs conceal them in rough nests or under debris, and one or other of the parents usually stands guard until they hatch. The eggs of this second kind, too, have some albuminous jelly around them—a modest safeguard against injury or a shortage of water which helps to free them from complete dependence on standing water.

Like frogs and toads, salamanders and newts usually mate at night to avoid the sun's heat. But unlike frogs they are voiceless—or nearly so. Pick one up and it will give a faint squeak or croak of protest—or possibly it is merely the air escaping from the creature's body. Because they lack the sound beacons of frogs, male salamanders must use another method to attract breeding females. You've probably guessed it—males and females find each other by means of their strong sense of smell.

Let us look at the courtship of the seven-inch-long California newt, *Taricha torosa*. (You may need flashlights to find your way through to the pond where breeding males gather first to await their brides.) Just as male frogs do, male salamanders emerge first from hibernation and move, usually, along streams to the breeding pool. When nights are damp they sometimes move overland. (Your flashlight won't disturb them at their nuptials, by the way; they are too intent on their courtship. In fact, less romantically, you wouldn't disturb them even if they were merely feeding. Occasionally you see salamanders in daylight—they are air breathers and if they are in the water they have to come up.)

As soon as a female appears she is encircled by excitedly swimming males. They swirl around her, rubbing and butting each other and waving their tails. One succeeds in straddling her from behind, clasping her with his arms below her armpits. The two perform a locked swimming dance; the male rubs his cloaca (vent) over her back and strokes her snout with his chin. In his chin is a scent gland which helps to persuade the female to accept his suit. Finally he leaves her and discharges packages of sperm. Called *spermatophores,* they are shaped rather like miniature tree stumps or collar studs. They have a broad, sticky base and a white stalk on which there is a top layer with millions of spermatozoa. If the female has been successfully wooed, she picks up one of the spermatophores with her cloaca. Inside her body the tiny sperms swim upward to her egg tract, and thus her eggs are fertilized internally.

The courting techniques of the male California newts rather resemble the breeding frogs we looked at earlier—that is, they gather together to await the females. The techniques also resemble those of some birds that we'll examine later, and those of the European smooth newt (*Triturus vulgaris*) resemble some birds even more closely. (The smooth newt was described and given its scientific name for the first time by the famous naturalist Linnaeus in 1758.) For instance, the smooth-newt male acquires a lovely nuptial attire during the breeding season. His tail fins and his dorsal crest become enlarged, frilled and brilliantly red along the edges—a vivid contrast to his workaday black-spotted brown body. He has become, in the words of Doris M. Cochran, "one of the most delicately beautiful of all living creatures"—and the delight of anyone who has a male and female pair in a vivarium.

The European newt is terrestrial, but in the breeding season he takes to the water to court the females with displays as beautiful as those put on by some fishes such as guppies or Siamese fighting fish—or, again, some birds. He swims about with abrupt jerky movements until he finds a female. Then he follows her about, nuzzling her sides and her mouth, a newt's version of a kiss. She is usually not persuaded immediately—after all, some time must elapse before his courtship signals get through to her—and she swims away. The male, his excitement mounting, eventually swims ahead of her and shows himself in all his marriage finery. He shakes his tail so that the toothed crest undulates in a beautiful ripple of pulsing color, and he swings his body into sinuous curves. The display concluded, he swims down to the bed of the pond and discharges two or three spermatophores. If she is ready for mating, she swims down to one clump and takes it up with her cloaca. When she lays the fertilized eggs, she glues them either singly or in small clusters to the stalks of water plants or to stones.

One of the most widely distributed is the crested or warty newt (*Triturus cristatus*), which lives in much of Europe—except the far north and extreme south. This newt, like many others of the order Caudata, has a skin that exudes a sticky poison, which often protects the animal from predators. A fox or dog who seizes a crested newt in his mouth drops it quickly because of the burning pain the poison induces. (The California newt also exudes a similar alkaloid poison which deters predators.) Emerging from winter hibernation, the males grow strongly toothed crests in April. Mating follows swimming dances similar to those of the California newt. The female usually lays a single egg at a time and conceals it by wrapping it in a water leaf.

The metamorphosis of the crested-newt tadpole to newt follows much the same pattern in all those species that lay eggs in water. The tadpole that emerges from the egg is almost transparent. Just behind the head and on each side are three small, fleshy-looking protuberances which develop into pale-pink feathery gills. These are later absorbed and become interior lungs. The metamorphosis is much like that of frog larvae into froglets except that the front legs appear first, not the hind legs. The metamorphosis of the crested newt takes about five months, depending on the weather and the supply of food. But if the eggs hatch late in summer, metamorphosis will be delayed until the spring.

Sexual maturity is reached usually in about three years. Sexual maturity in other newts and salamanders varies from two to four years, depending on the species, on the temperature, and on the quantity of food.

Lungless salamanders, which breathe through their skins and the lining of their throats and mouths, comprise the largest family (Plethodontidae) of the Caudata, with about 150 species. Some of the family lay eggs on land; others produce live young who pass the larval stage in the female's oviduct. The egg-layers show the beginnings of parental care; the eggs are usually guarded by the female until they hatch. One interesting member of the family, the arboreal salamander, *Aneides lugubris,* found in the Pacific area of the United States, lives up to its name. Doris Cochran records one of the six-inch salamanders being found 60 feet up a tree in the nest of a red tree mouse. This salamander lives in tree stumps, decayed logs, rock walls, mine shafts and even in basements. Eggs have been found 30 feet up in hollow trees. They're guarded by their mother until they hatch.

Another member of this family is the brown scelerpes (*Hydromantes genei*). Growing to about four inches and found in the mountains of southeast France, Sardinia and Italy, these salamanders produce living young about one and a half inches long.

One of the most famous of the newts (*Salamandridae*) is the fire salamander (*Salamandra salamandra*) of Europe, whose young bypass the egg stage outside the mother's body. Fire salamanders mate on land, usually in July, and the larvae hatch and grow inside the female's body. The 10 to 40 young are born the following April or May with arms and legs well developed but with gills. Deposited in the water, they can swim immediately. When they have become salamanders and are about two and a half inches long, they go ashore to spend their lives there.

Fire salamanders were long believed to be born in the flames of fires. It's not difficult to imagine how this myth got its start. Our ancestors, no doubt, threw a log with a salamander inside onto a fire and later saw the creature crawling away from the flames. A salamander surrounded by flames was the crest of some of the kings of France.

By the way, a nonmythical oddity about this salamander is that when it's handled it gives off a smell like that of vanilla! So, too, does the common European toad.

The black or Alpine salamander (*Salamandra atra*), growing to about six inches and found in the European Alps, produces live young and sets something of a record for prolonged pregnancy—two to three years for black salamanders living up to 3000 feet, and up to four years for those living at 4800 to 5400 feet. Only two young are produced, and they are not larvae but fully developed adults, about one and a half to two inches long. Each of the two ovaries produces about 20 eggs, but only the first egg that hatches develops. The larvae feed on the other eggs—one feeding in the left oviduct and one in the right oviduct.

The largest of the Caudata are the giant Japanese salamanders (*Megalobatrachus japonicus*), growing to over five feet and feeding on fish, other salamanders, worms and insects. The first one brought to Europe in 1829 by its discoverer, Philippe François de Siebold, lived for a further 52 years.

The Chinese giant salamander (*M. davidianus*) grows to about three and a half feet and was discovered for science by the famous Jesuit naturalist Père David. (We also owe to him the preservation in the world's zoos of Père David's deer). This salamander has beardlike barbels under its chin, and in the past many Chinese considered it a "man-fish."

Both of these giant salamanders lay about 500 eggs in long strings which are guarded by the male.

The North American representative is the hellbender

(*Cryptobranchus alleganiensis*), found in the eastern United States and growing to about two and a half feet. It is nocturnal and prefers running water. This creates a problem with egg-laying, which the female overcomes by pressing the long, sticky strings of eggs onto logs and stones at the bed of the river. The gelatinous casing hardens on contact with the water, so the eggs stay put until the young hatch, some 70 to 84 days later.

Mud puppies (which don't bark) and olms, belonging to the family Proteidae, are the Peter Pans of the order Caudata. They never pass beyond the larval stage and so never lose their gills. They breed without ever becoming truly adult.

Mud puppies (*Necturus maculosus*), which grow to about 14 inches, wear gay dark-red "plumes"—their feathery gills. The female usually lays light-yellow eggs, which she attaches singly to submerged logs or stones. Occasionally a female produces live young—tiny, fully developed adults.

Olms (*Proteus anguinas*), which are close relatives of mud puppies but separated by the wide Atlantic Ocean, have similar breeding habits. But as olms live and breed in total darkness in caves in Yugoslavia, little was known until recently about these mysterious long, eellike white creatures with brilliant carmine gills. (The bright red comes from the concentration of blood vessels in the gills.) In Yugoslavia they're still the subject of many legends. With their tiny hands and feet, they have a rough resemblance to a human embryo, and the peasants call them "human fish." At times of heavy flooding, olms are washed out of the caves into the lakes and fields, a phenomenon which puzzled both simple peasants and expert biologists for many years.

In an attempt to discover the life-cycle of olms, Professor Albert Vandel, Director of the Underground Laboratory of the Centre National de La Recherche Scientifique, Moulis, France, and his coworkers established a breeding colony of olms in 1952. Since that time many of the secrets of the creatures have been revealed. Professor Vandel found that his blind olms must have running water and could be maintained in light with no apparent injury. But eggs and embryos were so sensitive that they were killed by the light from a 100-watt bulb.

In Professor Vandel's controlled environment, both the male and female perform nuptial water dances, after which the male deposits spermatophores, which the female takes into her body. The eggs (20 to 60) are laid over a period of up to three weeks and placed on stones to which they adhere. Both parents guard the eggs and maintain a circulation of water by waving their tails. This prevents mud from settling on the mucilage that keeps the eggs adhering to the stones. The eggs hatch from 100 to 120 days after being laid. The water temperature is about 55° F., corresponding to that of the water in the Yugoslavian caves. The eggs laid are well furnished with yolk; after hatching, it takes about a month for the young olm to absorb it completely.

The young, Professor Vandel found, took eight to ten years to reach a length of six inches and sexual maturity. (This slow growth is typical of most cave-dwelling animals.) Their life expectancy is believed to be about 30 years. Although olms are hardy animals, they had to be well fed to reproduce. But, if necessary, Professor Vandel proved, olms can go without food for two years without any apparent ill effects. The only obvious effect was that their tails got shorter.

Mud puppies and olms are called *neotenic* by zoologists. It's an interesting word formed from two Greek words, *neos* = young, and *teinein* = to stretch or extend. Thus in these two Peter Pans you have creatures which retain larval characteristics beyond the normal period and are able to reproduce.

This Peter Pan characteristic of breeding in the larval stage is also possessed by a number of North American salamanders, including the famous Mexican axolotls. Some, such as the tiger salamander (*Ambystoma tigrinum*), metamorphose from larva to salamander in a few weeks or months in the eastern parts of the United States. Thereafter they are terrestrial and, growing to 13 inches, are the largest terrestrial salamanders in the world. But in the western states, such as Colorado and Wyoming, and in Mexico, metamorphosis often fails to take place. The same thing happens with some others living in Mexico belonging to the genus *Ambystoma.* It is believed that the shortage of iodine in the lakes of Colorado, and Wyoming and Mexico is responsible for this Peter Panism. (Axolotls, as experiments have shown, can be stimulated into becoming adults by being fed on thyroid extract.) It is therefore assumed that some factor in the natural environment—such as a shortage of iodine—prevents the release of the thyroid hormone into the circulation. Cold also apparently inhibits the release of the hormone. Some salamanders, such as tiger salamanders, fail to metamorphose at high altitudes in the Rockies. Others in warmer waters also fail to undergo metamorphosis—perhaps due to some inherited defect. As we've seen earlier, olms in cold water and mud puppies in much warmer water stay resolutely neotenic for the whole of their lives.

We have looked at the three orders of living amphibians and their eggs, and we have seen that their invasion of the land has been only partly successful. They are still tied to freshwater or to damp places and have won bridgeheads only. The further conquest of the land will be made by the reptiles, who are the subject of the next few chapters.

22

THE REPTILIAN EGG

THE LAND EGG or reptilian egg, as it is also called, has a very special place in the story of life as it evolved on earth. The land egg is one of nature's greatest innovations. It made possible the conquest of the land, first by reptiles and then by birds and mammals.

If the land egg had not developed, the land would have remained largely empty except for plants, invertebrate life and amphibians. As we have seen, amphibians are not strictly land animals; they cannot venture far from water, and most must return to the water to lay their soft, jelly-coated eggs.

Some time after the first amphibians developed, evolution took a decisive leap forward. The first reptiles invaded the land. (The word "reptile" is derived from the Latin root *repere*, to creep or crawl.) These first reptiles, which had evolved from the amphibians, were able to do so because they had acquired an egg that could be laid and incubated on land. This land or reptilian egg was much more complicated than the simple amphibian egg. The water cradled and protected the amphibian egg against mechanical injury and desiccation. The developing amphibian got its oxygen and most of its food from the water, and its waste matter was discharged into the water.

A land egg if it was to be successful had to provide everything the water had.

Let us look closely at the extraordinary solution—the land or *amniotic* egg, as biologists often call it. The unfertilized hen's egg in your refrigerator embodies the features we'll examine. Enclosed in the calcareous shell is a rich supply of food—the yolk—which, in a fertilized egg, is connected to the digestive tract of the embryo. (A nineteenth-century writer, Wilhelm Bölsche, compared the yolk in the eggs of reptiles and birds to a large piece of thickly buttered bread on which the embryo could thrive.) Enclosing the developing embryo is a large sac, the amnion, which is filled with liquid and protects the embryo from injury and desiccation. The amnion is thus the embryo's own private pond. At the back end of the embryo is a tube and a sac, the allantois, which functions both as a bladder for waste matter and as a lung. Enclosing the amnion is a membrane charged with blood vessels, which takes in oxygen and discharges carbon dioxide through the porous calcareous shell which encloses the egg. Another sac contains egg white (albumen). Enclosing everything inside the shell is yet another membrane, the chorion. (The origin of "amnion" is from the Greek word for the bowl in which the blood of sacrificed animals was caught; "allantois" is derived from the Greek for "sausage," a fairly accurate description of this membrane's shape.)

The shell is porous in a special way—it lets gases in and out but sheds a reasonable amount of water. However, if you submerge a developing egg in water, its embryo will surely drown.

Thus, with its own food supply (yolk sac), its private pond

(amnion), waste disposal and lung (allantois) and protective shell, the tiny reptile was freed from its dependence on water, and the conquest of the land could be attempted.

All reptile eggs are rich in yolk, the "thickly buttered bread" for the young that emerge in most instances as almost miniature replicas of their parents and able to fend for themselves.

Just as in the eggs of birds, the amount of yolk varies considerably from species to species—ranging up to 50 percent or more by weight.

The eggs of most reptiles are more or less oval in shape; often they are elongated with rounded ends. Some, such as the eggs of marine turtles and some freshwater turtles, are round like table tennis balls—and about the same size.

Most reptile eggs are white or yellowish in color. The eggs of the leatherback turtle, *Dermochelys coriacea*, sometimes have green flecks. Reptile eggs don't have the wide range of coloration that helps to camouflage—and protect—the eggs of birds. "Since most reptiles bury or hide their eggs, such colour could have little purpose," Dr. Angus Bellairs points out in *The Life of Reptiles*.

Some reptiles such as crocodileans (crocodiles, all alligators, gavials, etc.) lay eggs with hard shells, composed, like those of birds, of calcium salts. Other reptiles that lay eggs with hard shells include land-dwelling tortoises and some freshwater species of turtles and tortoises. Hard-shell eggs, too, are laid by most geckoes. They are the exception among the order Squamata (snakes and lizards), where all other species lay eggs with leathery or parchment-like shells. The shells contain little calcium and feel soft and rubbery when you handle them.

The reptilian egg is, in fact, a blueprint for the eggs of mammals and for live birth. In the more complex mammalian egg, one of the membranes inherited from the ancient reptilian egg, the chorion, comes into close contact with the wall of the uterus to form the placenta. Linked with the blood vessels of the allantois, the placenta supplies oxygen and nutriment to the developing embryo and carries away carbon dioxide and waste materials.

No one knows for certain how this wonderful egg evolved. One theory is that the shelled egg with its private pond was developed because of the great hazards experienced in the water by the larvae—"tadpoles"—of the amphibians. During the Carboniferous period, which succeeded the Devonian and preceded the Permian, waters were thick with the sharp teeth and clashing jaws of predators. A shelled egg laid on land in which the embryo could develop and emerge as a miniature replica of its parent offered much greater chances of survival.

Another theory is that the first land eggs developed among amphibians living in fast-running mountain streams where fertilization in water was highly uncertain or even impossible.

Whatever the reason, or reasons, there was great pressure on the amphibians to evolve this wonderful egg.

The ancient reptiles had to make a further innovation to succeed in the conquest of the land. A female frog lays her eggs in water. Simultaneously, the male who is clinging to her back releases his sperm. Fertilization thus takes place in the water. The eggs of the primitive amphibians were also fertilized in the water.

But, for the obvious reason that the reptilian egg had a tough outer shell when laid, fertilization among reptiles had to take place within the female's body. In the primitive reptiles the transfer of sperm from the male to the female was probably achieved by bringing the cloacas into contact. This is also the method used by most birds. (Most male birds have no penises—the exceptions are ducks, ratites,* tinamous, curas-

* Ostriches and other large, flightless birds.

sows and a few others. This bird method is that of the most primitive of all living reptiles, the tuatara of New Zealand, which has no penis.)

In all modern reptiles the male has an organ which, like an inverted finger on a glove, is kept inside except when it is needed for fertilization, at which time it springs out. (This glove-finger type of penis is described in more detail later.)

In reptiles—and in birds and mammals, too—the ovum or egg is one of a large number of ova enclosed in the membranous follicles (Latin for "small bags") of the ovaries. These follicles may number as many as 25,000 in some reptiles and birds, but only a few ripen each breeding season. The ripe ovum accumulates a large yolk and in time bursts from its follicle and enters the oviduct of the reptiles. If sperm are present, the ovum will be fertilized while moving down the oviduct. Here it also acquires a deposit of albumen. Descending into the uterus, the ovum acquires its various membranes and finally the protective shell.

To conquer the land, reptiles had also to acquire a dry scaly skin which conserved moisture. Having done so, and having developed the wonderful land egg, the stage was set for the reptiles' conquest of the surface of the earth. In the Carboniferous and Permian periods the reptiles expanded greatly, and during the Mesozoic (Middle Life) era that followed, their dominance was so complete that it has come to be called the Age of Reptiles. During this era of about 145 million years, the reptiles were the lords of creation. This is an immensity of time that all of us find hard to comprehend. We can perhaps grasp it better if we imagine the life of the earth (some 4.6 billion years) as a 30-story building. The time when the reptiles were the masters of the earth would occupy the top half of the twenty-ninth and the bottom half of the thirtieth floors—that is, about 14 feet altogether. About 7 feet from the ceiling would represent the last 70 million years. Man's history on the earth would occupy a few inches in this space below the ceiling. And our *recorded* history would be the top two coats of paint!

For most of us the Age of Reptiles says, "Dinosaurs!"— the most spectacular and awesome of all the reptiles. Incidentally, the great anatomist of the last century, Richard Owen, coined the name "dinosaur" from two Greek words *deinos* = great or terrible, and *sauros* = lizard. The last part is not too accurate; the dinosaurs were not lizards—lizards developed later. (We tend to think of dinosaurs as gigantic creatures. Some were, of course, but they ranged from creatures the size of chickens to giants exceeded only by whales. Most people think of these cold-blooded creatures as slow-moving, sluggish animals, but many were quite agile, and some two-legged ones probably had a top speed of 30 miles per hour.)

This book is only concerned with their eggs. Appropriately, the oldest land egg we know of is a dinosaur egg—a fossil egg found in the lower Permian deposits in Texas and thus about 225 million years old. (A number of primitive reptiles existed at that time; we don't know which of them laid the oldest known egg. There were earlier reptilian eggs, but we've found no fossils as yet.) You'll probably be disappointed to learn it's a rather small egg—only two and three-quarters of an inch long and slightly larger than an average domestic hen's egg. Larger fossil dinosaur eggs, oblong and about ten inches long, have been found in Provence, France. Probably, for reasons I'll explain later, the largest dinosaur eggs were never as large as those laid by the extinct elephant bird (*Aepyornis titan*) of Malagasy (Madagascar), which were thirteen by nine and a half inches. This probably astonishes you just as it astonished a friend of mine. When he learned I was writing this book, he did a calculation of the supposed weight of a dinosaur egg, basing his calculations on the fact that a female ostrich lays a three-pound egg, which is about a sixtieth of her body weight. My friend estimated that a *Brontosaurus*, up to 70 feet long and weighing 30 metric tons, would have laid an egg that weighed half a ton or 1000 pounds! He'd have gotten even more remarkable figures with *Brachiosaurus,* which weighed from 50 to 80 tons!

Alas, his calculations are contrary to the laws of physics. Very large eggs would have to have extremely thick shells to withstand the pressure of their internal fluids, and this would have meant the baby dinosaur could never get out. The shell of the reptilian egg, while much stronger than that of an amphibian egg, is more fragile than that of birds; probably the biggest dinosaur egg ever was about the size of those found in Provence.

One of the most exciting finds of dinosaur eggs was in the 1920s when an expedition from the American Museum of Natural History, led by Roy Chapman Andrews, found several "nests" of fossilized eggs in Outer Mongolia. The first find was of three petrified eggs sticking out of a sandstone ledge. Under the shelf were others—13 eggs, each about eight inches long, in two layers, their rounded ends pointing inwards. An even more exciting discovery followed—two of the eggs contained the tiny petrified embryos of unhatched baby dinosaurs. These eggs, about eight inches long by three inches across, were identified as those of *Protoceratops,* the first of the horned dinosaurs of Cretaceous times, which began about 135 million years ago. *Protoceratops* grew up to seven feet, weighed 900 pounds, had jaws that looked like a parrot's beak and had an upturned "ruff" of bone on the back of its head. It and its fellows were vegetarians.

More eggs of *Protoceratops* were discovered by the same expedition. Some nests had 30 to 35 eggs in circles and in layers. And, excitingly, on top of one nest was the petrified skeleton of a small toothless dinosaur that preyed on dinosaur eggs. We can imagine that perhaps a huge sandstorm overwhelmed the predator and the eggs it was raiding and buried them both for over 70 million years.

Nine different kinds of ancient eggs have been found in Provence. At one of the evocative sites you can project yourself back for millions of years to see how a female dinosaur laid eggs. She deposited a few eggs and then took a step forward and laid a few more. She did this 15 to 20 times. At Montarnaud near Montpellier there are long, winding rows of large, round eggs, looking like so many cannonballs.

We do not know if some of the prehistoric reptiles guarded their eggs. Probably most took no further interest after they had laid them, just as do many living reptiles. Almost certainly none took any care of their young after they hatched; the young had to fend for themselves just as baby turtles do today.

In the chapters that follow we'll look at some of the land eggs laid by living creatures. It is perhaps appropriate that we should begin with the reptiles, those remnants of the great reptiles of the Mesozoic era.

The dinosaurs became extinct, as everyone knows. You should not think of this as happening "suddenly"—that is, in the space of a few thousand years, or even a few tens of thousands of years. Like Charles II, the reptiles were "an unconscionable time dying." Their numbers probably declined very slowly over a period of millions of years, until they were extinct and had given way to the more efficient (that is, adaptable) mammals.

But not all the reptile orders became extinct. Four survived into our times—a sad decline from the days of reptile greatness in Mesozoic times when there were about 17 orders and suborders with many thousands of species. The orders of reptiles we know are:

Rhynchocephalia	1 species of tuatara
Crocodilia	25 species of alligators
	crocodiles, caymans, gavials
Chelonia	about 350 species of turtles
Squamata	about 2700 species of snakes and
	3000 species of lizards.

The tuatara survived because of a happy accident of geological isolation. In its New Zealand island home, this small lizardlike creature faced no competition from either more efficient mammals or reptiles.

Crocodilians persisted in tropical habitats because they were well-armed and armored predators; the turtles were even better protected. The Squamata developed in highly special ways from small-legged ancestors; they survived in competition with mammals by living secluded and even secretive lives among foliage and rocks; and some survived in part by acquiring venom to kill their prey and deter their enemies.

What also probably helped these reptiles to survive was the practice some acquired of burying their eggs—and in a few cases of caring for their young. Burying the eggs not only provided the warmth and moisture essential for their incubation but also some degree of protection.

About all the reptiles clings the nostalgia of past glories. They are the melancholy remnants of a great and defeated army. Three of the orders—Crocodilia, Chelonia and the solitary Rhynchocephalia—are truly "living fossils."

23

TURTLES

ON THIS early summer's night the long rollers of the Pacific are breaking and foaming up the North Queensland beach. As our eyes become accustomed to the darkness they pick up a rocklike shape. The "rock" moves with the waves, rolling, submerging and reappearing, each time closer to the shore. We realize that it is a female green turtle (*Chelonia mydas*) coming ashore on the near-full tide to lay her eggs. She is about three feet long and weighs about 250 pounds. She has come from perhaps 1000 miles away to mate in the waves offshore.

Now she rides with the waves, heavy with her eggs, until she is stranded. Then the struggle begins up the sloping beach. With her powerful flippers she "swims" laboriously through the sand, winning slow yard upon slow yard. On land she is a clumsy cripple, but once in the water she is as agile as a seal; her flipper action is like the up-and-down action of a bird's wings so that she quite truthfully "flies" through the water. She can cover a hundred yards in under ten seconds, faster than all but Olympic athletes can run.

Her barnacle-encrusted carapace (shell) looms larger in the gloom as she climbs up the spinifex-clad dune. It's no accident she's come to this spot. On several nights she has come ashore to search for just such a place as this—on the crest of a dune, well above the high-water mark, with a steepish slope down to the beach. The sand must have sufficient grass or other roots to prevent sand falling back into the nesting chamber while she digs. And there must be some moisture in the sand which the eggs can absorb—freshwater only, because salt water is toxic to the eggs.

As we know, the first reptile developed one of nature's greatest inventions—the land egg containing its own pond. This "invention" is not complete with turtles—hence the need for the eggs to absorb some water from the sand.

On the top of the crest the green turtle noses about until she finally decides on a place. She begins to dig a pit, using first her front flippers and then her hind flippers to pile the growing heap of sand behind her. After nearly an hour's work the pit is about four feet wide and about two feet deep. Now she can start forming it into her nesting chamber. Using one hind flipper and then the other, she curls the ends of her flippers, cupping them like hands, and lifts the sand out, slowly and methodically, about a cupful at a time. Forty minutes later she has a bottle-shaped hole about two feet deep. She has moved well over a cubic ton of sand! While digging she has paused frequently and gasped.

Now she maneuvers her protruding tail into the hole and drops her round, moist, ping-ponglike eggs, one or two at a time. In this fashion she lays perhaps 20 eggs, then rests for a few seconds before resuming her egg-laying at the rate of about one a second. In this way she lays over a hundred eggs.

When she is finished, she scoops the sand back into the hole by dragging herself forward with her foreflippers. Sometimes, bulldozer fashion, she'll completely fill the hole with one long, slow push. When the hole is filled she scatters sand all over the site with her front flippers. She yaws her body about until the egg site is obscured and not easy to detect. (Sometimes she may move a few yards and scratch up the sand to make a false nesting site.)

Something like three hours have passed since she arrived at the site on the crest. Now, having laid and hidden her eggs, she makes her way back to the sea. She is weary now and takes frequent rests on her way. In the water, now very exhausted, she waits for a wave to sweep in and carry her back out from the beach. Her domed shell gleams suddenly in the spray, and then she is gone, back into the sea. This is her first visit this summer to the nesting beach. She may return as many as six times, laying in fresh nests 50 or more eggs at intervals of about a fortnight. (Caches may vary from 60 to 200, and in a season she may lay 350 to 500 eggs.) She will return perhaps in two or three years' time. Green turtles usually breed once in every three years. A few breed once in every two.

Like all turtles, her motherhood ceases with the laying of her eggs. The eggs she has laid on this night will take two to three months to hatch, depending on the warmth of the sand. During this time they may be raided by monitor lizards who nose out the nests, slash open the tough leathery shells with sawlike teeth and suck the contents.

Green turtles lay eggs on the beaches of all the tropic areas of the world. In all regions, lizards prey on them, and in Africa, mongooses frequently discover the succulent eggs with their high yolk content. (Fifty-five percent of a turtle egg is yolk, compared with about 35 percent of a domestic hen egg.) Rats, dogs and men, too, seek turtle eggs. Dogs and rats locate the nests by smell; in northern Australia, aborigines plunge steel rods into the sand. When the tip comes up wet they know they have found the nest.

Baby turtles run a desperate paddlefooted race with

death to reach the sea. At night, when most make their run for the water, they're preyed on by crabs. During the day, hawks and gulls hover, swooping to attack the tiny turtles with their soft shells. (Later they'll grow hard.) In the sea, fish wait to attack. As a result, few baby turtles reach maturity to return to this beach, seven years from now, to lay their eggs.

In tropical waters estuarine or saltwater crocodiles (*Crocodylus porosus*), too, lurk in wait for baby turtles—and for their mothers. The stomach of a large crocodile killed not long ago at Arnhem Land, northern Australia, contained a female

FIG. 7. *Hawksbill turtles crawling back to the ocean after digging nests and laying eggs. The shells are encrusted with barnacles. The time, early morning.*

turtle that was about to produce a clutch of eggs.

All seven species of marine turtles, divided into two families, are found throughout the warm waters of the world, and all return to the beaches to lay their eggs.

Green turtles are mainly vegetarians and can attain a weight of 850 pounds, but today, because of our hunting of the creatures, a turtle weighing about a third of this is a comparatively large specimen.

Largest of the sea turtles is the luth or leatherback turtle

(*Dermochelys coriacea*). The carapace of this fish-eating turtle is covered by a leathery skin that unites a mosaic of small separate bones. The leatherback may become eight feet long and weigh up to 1500 pounds. Among reptiles it is second only to the estuarine crocodile, which may weigh more than two tons. The female leatherback lays up to 350 eggs at each nesting. Before laying, she dusts herself lightly with sand, probably in an attempt at camouflage. While laying she "weeps" copiously. This habit is believed to help in keeping the sand out of her eyes. Many of the marine turtles "weep" in this way.

Another sea turtle, the hawksbill turtle (*Eretmochelys imbricata*), was the source of the tortoiseshell of commerce. (See Figure 7.) The dermal plates of the hawksbill used to be fused together by steam under pressure to form tortoiseshell. Today synthetic substitutes have largely replaced tortoiseshell.

Another large marine turtle is the loggerhead (*Caretta caretta*).

Many of the sea turtles appear to migrate, apparently in search of food and suitable nesting beaches. We know most about the green turtles. They feed on marine plants that grow in protected waters, and for breeding they need broad beaches on which unchecked waves can break. Some green turtles commute between their grazing grounds on the coast of Brazil and lonely Ascension Island, 1400 miles away in the South Atlantic. Caribbean green turtles travel up to 1200 miles from the single nesting beach in the Western Caribbean—at Tortuguero in Costa Rica. So far as we know, the green turtle we watched nesting on the Queensland beach doesn't migrate in the way its American cousins do. But, nonetheless, they range widely; some observed in Queensland have been seen later 1000 miles away in the empty Pacific.

The navigational skills of these sea turtles in the vast open stretches rival those of many birds. They are believed to take their bearings from the sun and the stars. Such feats of navigation interest the U.S. Navy, which, in attempts to track the turtles and unravel their mystery, has fitted some with balloon-borne transmitters. The famous American expert, Dr. Archie Carr, of the Chair of Zoology, University of California, wants satellites used to check on the journeys of green turtles. The turtles would carry some transmitters whose signals would be recorded by the orbiting satellites.

It's little wonder the U.S. Navy and zoologists are interested. It is something of a navigation miracle. Think for a moment of the baby turtle who hatches on Ascension Island, a comparative speck of 34 square miles in hundreds of miles of apparently featureless ocean. The baby reptile sees the world for the first time from the bottom of a hole. It moves its head from side to side for a minute or two. And then it heads directly for the sea. Night or day, its "radar" works unerringly, and within half an hour or so it is in the water and steering a course for Brazil and its ancestral grazing grounds. Then, when it matures sexually, it heads back for tiny Ascension. Tagging by Professor Carr has established that some may even land on the very beach where they were hatched, and some have come to their home beach repeatedly. It's clever navigation because even a very small error would make them miss Ascension by many miles.

On at least one occasion a seagoing green turtle has saved a man's life. In 1969 Chung Nam Kim, a sailor, fell off a Liberian ship off Nicaragua and after 15 hours of swimming was giving up hope of surviving. Along came a giant green turtle. He climbed onto its back and rode there for two hours until he saw the Swedish ship *Citadel* whose astonished crew hauled him on board.

Other turtles include the soft-shelled turtles found in North America, Africa, the Asiatic region and New Guinea. They have a flexible shell covered with soft skin. Another large group include the side-necked and snake-necked turtles found in South America, Madagascar and New Guinea. They draw their heads and necks into the recess of the shell by bending the neck laterally to one side.

The remaining turtles have hard shells and can draw their heads straight back into their recess. They are found on all continents (except Australia).

When is a turtle a tortoise, or vice versa? Strictly, "tortoise" is a name generally reserved for a group of land turtles with stumplike feet, which usually feed on vegetation. Popular usage is confusiong and contradictory. For instance, as we've seen, the sea-dwelling hawksbill turtle was the source of tortoiseshell. The box turtle (*Terrapene carolina*) of North America lives almost entirely on land and rarely takes to the water. Those of you who have kept this popular pet know it likes meat as well as lettuce leaves. (In the wilds, box turtles snack on snails, slugs, worms and insects.) Dr. Archie Carr had one that regularly fed on hamburger and would, much as a pet dog might, waddle to the refrigerator and wait for Carr to take the hint.

Some tortoises grow to immense size. Those on the Galapagos Islands sometimes weigh as much as 500 pounds, and there's a record of one killed on Isabela that weighed 560 pounds. When just hatched they weigh about three ounces! (*Galapagos*, by the way, is Spanish for "tortoise"; the 14 islands are about 600 miles west of the coast of Ecuador and astride the equator.)

Other large tortoises, some up to 500 pounds or more, live on some of the islands of the Indian Ocean—the Seychelles, Rodriguez, Aldabra and Mauritius. The nearly round eggs of these giants and those on Galapagos are about as big as tennis balls.

In all there are about 350 species of turtles and tortoises, and what all have in common, turtles and tortoises, is the habit of laying shell eggs buried in the ground and left to hatch of their own accord.

Because baby tortoises generally have fewer enemies than baby marine turtles or are better camouflaged, tortoises need lay fewer eggs in order to maintain their numbers. For instance, the box turtle lays from one to eight eggs in a shallow nest dug in soft soil. And the African soft tortoise is thought to lay one egg only. The box turtle lays her eggs usually in late June, and the young hatch out two or three months later. Their shells blend well with the soil and vegetation, which is just as well because at first the baby turtles, which are about the size of a quarter, cannot withdraw completely into their armored houses. Some of their enemies are coyotes, dogs, badgers, skunks, raccoons and birds of prey.

The eggs of tortoises are usually white, hard-shelled and almost spherical.

Age, too, determines how many eggs are laid; young females lay fewer eggs than older ones. Some tortoises moisten the pit with water from their bodies before laying. All cover their eggs and smooth the soil to leave as little trace as possible. Some even lay an egg or two in isolated pits a foot or so from the main clutch. These "decoy" eggs probably help to divert raiders.

Adult tortoises with their tough armor are relatively invulnerable. In some parts of Europe eagles are reputed to carry a tortoise high in the air and drop it on a rock so its shell splits. It's probably a tall story. You will recall that the Greek dramatist Aeschylus is said to have been killed when an eagle mistook his bald head for a rock and dropped a turtle on it!

Incubation periods of some tortoises can be very long. For instance, some eggs of the Australian broad-shelled tortoise (*Chelodina expansa*) took 360 days to incubate.

Baby turtles and tortoises have a caruncle or horny projection on the tip of the snout; they use this "tooth" to peck their way out of the leathery shells and later lose it. Turtles and tortoises have no teeth.

Before a female turtle or tortoise can lay her eggs, there

must be a mating with a male. Like all birds and reptiles, the egg must be fertilized in the female's body and before it has acquired a protective shell. The male organ in turtles and tortoises is a distensible, grooved penis, which is attached to the outside wall of the cloaca. It increases in size by turning itself inside out. (Like birds, reptiles have only one vent for sexual and excretory purposes.) The female has two ovaries connected to the cloaca by an oviduct. The opening is near the base of the tail.

Usually, in nature, eggs are fertilized just before being laid. But among turtles and other reptiles the sperm may be stored for months or even years to fertilize eggs when they're formed. Not infrequently a tortoise you buy from a pet shop may lay fertile eggs 12 months or more after she could have mated. Record instances are of a diamondback terrapin (*Malaclemys terrapin*) and a box tortoise (*Terrapene carolina*) which laid fertile eggs four years after fertilization. The diamondback terrapin is a North American member of an almost worldwide family (*Emydidae*) found mainly in the freshwater and called variously turtles and tortoises. Sperm storage has been recorded in other reptiles, most notably for four and a half years in an indigo snake (*Drymarchon corais*) and six and a half years in another colubrid, *Leptodeira*. Dr. Angus Bellairs suggests in *The Life of Reptiles* that in the instances of the two snakes parthenogenesis "must be considered" as a possibility.

Before they can mate, of course, male and female turtles must find each other. Courtship is relatively silent because, like other reptiles, they have no vocal chords. (The passage in the Bible, "And the voice of the turtle is heard in the land," does not refer to turtles but to doves.) Some by expelling air produce grunts, moans, roars and hisses. The zoologist Karl Escherich claimed that a Moorish tortoise croons a love song that resembles the distant cries of a baby.

Generally marine turtles court with gestures made with their flippers. A courting pair perform a kind of swimming dance. Many tortoises, too, signal their love to each other with quivering gestures of their forefeet. Or the male may bob his head and bite harmlessly at the female's shell. Males of other species, again, may bump her shell or scrape it with their feet in order to press their suit. Males sometimes fight each other. They begin with ritual bobbing of their heads and try to ram each other like miniature tanks. Some try to bite each other. Others go in for pushing matches, which continue until one retires or is thrown on his back. These mating battles are noisy affairs.

The western painted turtle (*Chrystemys picta*), whose upper shell carries red and yellow markings, gently strokes his partner's face with the long nails on his forefeet. Females do the wooing among the spotted turtles, which have yellow markings on their shells; the females have saffron-yellow faces that are love lures to males.

Female Galapagos tortoises fight with each other for nesting sites. They will continue scrapping even while gulls prey on the eggs the defending owner has laid.

Turtles and tortoises are the "squares" of the animal world. They differ very little from their ancestors of 200 million years ago. The first turtles were contemporaries of the earliest dinosaurs—at a time when there were no birds or mammals on the earth. The dinosaurs developed to dominate all the earth for about 120 million years and in time passed away. The turtles survived mainly because they developed defensive armor into which the head, limbs and tail could be wholly or partially withdrawn.

Turtle defense reaches its peak in turtles such as the box turtle and some mud turtles. The underside of the box turtle's shell, called the "plastron," hinges in the middle. When the turtle feels endangered, it withdraws inside and pulls both ends so tight up against the carapace or upper shell that you can't get a knife blade through the crack at either end. The turtle "boxes" itself in so effectively that dogs are frequently baffled and will turn the shell over and over looking for the creature they saw only a few seconds earlier.

As we have seen, all reptiles are descended from the stem reptiles. We don't know the precise line of descent from the stem reptiles, but the ancestor of today's turtles could have been a small reptile found in fossil form in the 250-million-year-old Permian rocks of South Africa, which paleontologists called *Eunotosaurus* ("stout-backed reptile"). This four-inch-long reptile had broadly expanded ribs, which, as Dr. A. S. Romer says vividly in *Man and the Vertebrates*, gave it "somewhat the appearance of an umbrella plant" and suggested the beginning of the typical shell. Eight pairs of the ribs are expanded into broad bony plates. It presumably had a long tail—this was missing—and a thick neck and head that it was unable to retract. Other early turtle fossils have been found in Triassic limestone deposits. These turtles were presumably unable to withdraw their heads and necks into their shells, but otherwise they were substantially turtles as we know them today.

The ribs of the ancestral turtles developed into the armor that we know today. Thus the carapace consists of modified ribs and vertebrae covered with a thin sheet of horny substance. It is true to say that a turtle lives "inside its ribs."

Ancestral turtles had teeth but lost them in the course of evolution—just as birds, also descended from reptiles, have done.

Some prehistoric turtles were very large. The fossil remains of one North American turtle, *Archelon*, were found in Kansas. This gigantic turtle was nearly 11 feet long and must have weighed at least three tons. It had a large, parrotlike beak and lived on shellfish in the shallower seas of the then

American continent. Huge as it was, *Archelon* had its preying enemies. One fossil skeleton has the right-hand flipper bitten off well above the heel—possibly by a giant mosasaur (Meuse lizard). Mosasaurs were marine lizards up to 30 feet long with long, powerful tails and long necks.

Turtles reversed their evolutionary steps, as it were. Their amphibian reptilian ancestors left the water; turtles returned to it.

All turtles and tortoises are long-lived—something which has no doubt assisted their survival from the Age of Reptiles. A box turtle will live for 40 to 50 years—one is said to have lived for 130 years. A European common tortoise (*Testudo graeca*) lived in captivity for nearly 40 years. Another land tortoise, *Testudo sumeirei*, was brought to Mauritius in 1766 and lived for 152 years—until 1918, when it was accidentally killed. It was believed to have been between 200 and 250 years old. Another lived for 120 years in the barracks at Port Louis in Mauritius. And yet another land tortoise, one that had been a companion of Napoleon in his exile on St. Helena, was reported not long ago as still living—150 years later.

A long-lived Galapagos tortoise died on the royal-palace grounds in Tonga in 1966. Legend has it that this tortoise—and a companion, which died some years earlier—had been brought there by Captain James Cook on his third voyage, 1776–78. This would give the male tortoise a known life of nearly 190 years. The eminent Cook authority, Professor J. C. Beaglehole, thinks the legend is inaccurate and that the tortoises were left in Tonga much later. They could have been brought from Galapagos by other seafarers such as American whalers. What is reasonably certain is that one tortoise lived on Tonga for at least 100 years.

Another long-lived tortoise is Jonathan, who was taken from the Seychelles Islands to St. Helena in 1825, only four years after Napoleon's death there. Jonathan lives on the grounds of Government House and disrupts games of croquet by sitting on the balls. He is at least 160 years old and possibly even 170 because he was an adult of perhaps 20 to 30 years when shipped to St. Helena. Jonathan arrived with a bride, but she died over 100 years ago; one account says Jonathan accidentally pushed her over a cliff. In 1969 the governor of St. Helena, Sir Dermod Murphy, took pity on Jonathan's plight and dispatched a mission to the Seychelles to find another bride for Jonathan.

Despite this singular act of kindness, the fact remains that man is the worst enemy of turtles. The most endangered turtle is probably the green turtle, the source of the incomparable gourmet delicacy, clear green-turtle soup, made from the flesh of the calipee or cartilage of the lower shell. Green-turtle meat and eggs are also esteemed. Most countries today protect the creatures, but legislation is difficult to enforce on thousands of miles of beaches in the warmer seas of the world. The diamondback terrapin, the North American turtle of tidal waters, was once in great demand as a delicacy (terrapin stew). In 1920 when the vogue was at its peak they sold for $90 a dozen. Today, fortunately for the diamondback, the gastronomic taste in the United States is for the snapping and soft-shelled turtles.

Strenuous efforts are being made today to preserve the giant Galapagos tortoises, which are threatened with extinction. There were once so many that it was said a man could walk a hundred yards on their closely congregated shells without setting a foot on the ground! "They are so extraordinarily large and fat, and so sweet, that no pullet eats more pleasantly," declared William Dampier, the English pirate, in 1684. Hundreds of shells mark the sites of sailors' camps since Dampier's time. In the short space of 30 years from 1830 onward, American whalers are believed to have slaughtered about ten million Galapagos tortoises for food. Their numbers had shrunk so dangerously by 1928 that the New York Zoological Society sent an expedition to catch tortoises and move them to other places because it was feared they would soon be extinct on Galapagos. Some were taken to Florida, Bermuda and Hawaii. An important step in their conservation came in 1959 when a UNESCO foundation established the Charles Darwin Research Station in the Galapagos Islands to help preserve their unique wild life.

Reptiles were once the dominant animals, but their long day waned; today man is the dominant animal. But turtles are still here. And they could still be here when man has left the scene. Aesop wrote more wisely than he knew in his fable of the hare and the tortoise. In the long story of evolution, too, the race isn't always won by the swift.

24

CROCODILES

PROFESSOR ARCHIE CARR was wading in a Florida stream seeking frogs when suddenly a female American alligator (*Alligator mississipiensis*) charged . belligerently halfway across the stream. Professor Carr retreated hastily. In the shallows he discerned about eight small alligators, up to 16 inches long and about two months old. The mother's rush to defend her offspring, Carr says, quickly took his mind off putting any of her young in his bag.

The British traveler Sir Robert Schomburgk tells of an even more touching instance of maternal affection in a South American cayman. An Indian of his expedition climbed out on a tree trunk overhanging a Brazilian river, slew one of a group of young caymans with an arrow and lifted it out of the water. Suddenly the mother broke the surface among the young caymans, roared loudly and tried to attack the party, which was in canoes. Schomburgk writes, in *Travels in British Guiana*:

The hitherto calm surface of the water suddenly became an agitated mass of billows, since the cayman lashed it continually with her crooked tail, and I must admit that the incredible boldness of the animal made my heart beat twice as fast. . . . After we had exhausted our supply of arrows, I thought it wise for us to retreat as cautiously as possible. Obstinately, the mother followed us as far as the shore; but she came no further, for on land the cayman is too timid to be dangerous.

Such maternal affection is an exception among reptiles, who in most instances behave like the turtles we observed in the previous chapter; they lay their eggs, bury them and leave the young to take their chances with predators. There is possibly some significance in the fact that the American alligators' relatives belong to an ancient reptilian stock that is more closely related to birds than other living reptiles.

This posthatching care shown by the American alligator (which has a range from Florida to Colombia) and a South American cayman is thought to be a characteristic of a few other crocodilians, including the Asian estuarine crocodile (*Crocodylus porosus*) and possibly the Nile crocodile (*C. niloticus*), which is found in all of Africa south of the Sahara to just north of the Transvaal. But evidence is conflicting; some naturalists claim that the estuarine-crocodile mother sheds her maternal role immediately after her eggs hatch and is liable to look upon her offspring as a tasty snack.

These three crocodilians guard their eggs while they're hatching. With them something else, too, has entered the picture—the broodiness so familiar to us in birds.

The American alligator labors strenuously at her nest building on the edge of a pool or swamp. She scrapes up leafmold, twigs and branches and even living vegetation (which she bites off) and mixes it all with mud to make a huge mound that may be 15 to 18 feet across and 3 feet high. This done, she digs a pit in the center and lays from 15 to 80 eggs about 3 inches long with hard shells. (All crocodilians lay hard-shelled eggs, unlike the soft, flexible shells of most snakes and lizards.) She covers them with more leafmold and settles down to wait for the eggs to incubate with the help of the heat generated by the decaying vegetable matter. Her way of enlisting decomposing vegetable matter to hatch her eggs resembles that of some remarkable birds, the mound builders of Southeast Asia and Australia, whom we'll meet in Chapter 36.

We can risk a guess that this method of building a raised mound was forced upon the American alligators' ancestors—it was necessary to keep the nest well above the damp floor of the swamp and to find enough heat to hatch the eggs. As opposed to this practice of nest building, the American crocodile (*C. acutus*) digs a pit in sun-warmed sand the way a green turtle does and leaves incubation to the sun's heat. In shady swamps, the sun's heat would not normally be enough to hatch the alligator's eggs.

For up to two months the mother American alligator guards her nest, never venturing far away. An intruder is met with a bellicose rush that looks businesslike enough to convince most animals or men. After about 60 days her vigil ends; her seven-inch-long babies are breaking out of the hard shells. A baby lizard or snake has a tiny egg tooth to cut its way free; a baby crocodilian, like a baby turtle, has a caruncle, a horny projection on the tip of its snout, which it later sheds. The baby alligator takes several hardworking hours to free itself of the shell. When it does, it is trapped beneath the mud-plastered mound, which has set hard. The baby alligator and its brothers and sisters begin squeaking, and their mother hears, waddles over to the mound and tears it open. (Nile crocodiles do the same for their offspring; their nests are often baked so hard by the African sun you can't tear the mass open with your hands. The urge to dig out her babies is so strong among Nile crocodiles that when some scientists built a fence around a nest, the excited mother smashed and tore it to pieces. A recent study by Dr. H. B. Cott has brought to light that females with hatched young often behave aggressively.) Baby alligators can fend for themselves immediately, but maternal care, usually in the "gator hole" the mother has dug, will last for two or three months.

A family of small alligators observed by Professor Carr croaked together much like a flotilla of ducklings. Young alligators are black with yellow bands—a form of protective coloring they lose as they grow older and less vulnerable.

The only other alligator in the order Crocodilia, the Chinese alligator (*Alligator sinensis*), also builds a nest. (It is not known whether it practices postnatal care.) Something of a dwarf among crocodilians, the Chinese alligator grows to a maximum length of 6 feet and a weight of 75 pounds only, but was, nonetheless, sufficiently impressive to inspire most of the stories of Eastern dragons. Marco Polo returned to Italy with tall stories of a huge serpent with forelegs, which emerged after dark to eat sailors! The Bronx Zoo in New York, which acquired some Chinese alligators in recent years, has found the creatures' fierceness to be much exaggerated except during the breeding season. When a zoologist, Peter Brazaitis, entered the breeding enclosure, the female was, as he relates in *Animal Kingdom,* well out of sight in the pool at the far end. Then, he says, "Suddenly she rushed at me and I was forced to beat a hasty retreat." The male was treated only a bit less fiercely when he decided the nesting material might be a suitable place to bask.

He approached the female and an attack seemed imminent as again her tail began to flick threateningly. She slowly began to advance as the male settled down. Instead of attacking, however, the female gently seized his head in her jaws and began to push him from the material. As he stirred she transferred her grasp to his midbody. A weak bite at the base of the tail made her intentions clear. The nesting material would be hers.

Maternal guardianship of the nest and later of the youngsters undoubtedly gives these crocodilians some advantages over their fellows—and other reptiles—as indicated by the fact that an American alligator need lay only from 15 to 80 eggs as against the 350 to 400 eggs laid by a green turtle. Nevertheless, mortality is high among newly hatched crocodilians. Animals, birds—and other crocodilians—prey on them as they head instinctively for the water, running a gauntlet as hazardous as that essayed by baby turtles. In the water predatory fish lie in wait. It is estimated that of a clutch of 40 to 60 eggs laid by an estuarine crocodile, only one newly hatched baby reaches sexual maturity.

Mortality for most crocodilians begins from the day the eggs are laid. Tasty, by our standards, these eggs are sought by men and animals. American President Theodore Roosevelt found 32 eggs in a female crocodile he shot, and had them made into an omelette which he declared was delicious!

All told there are 25 species in the order, which includes the family of true crocodiles such as the Nile crocodile and the estuarine crocodile of India, Asia and northern Australia, which is also called the saltwater or ocean-going crocodile because it has been observed in mid-ocean, hundreds of miles from land. The order also includes the family of alligators and caymans (also spelled caiman), which are restricted to Central and South America, and the family of gavials (also called garials) of India and the Malay region. The double name for the gavial began because a scientist's handwriting wasn't legible. The scientist who first wrote about this crocodilian intended to give the genus the name of *Garialis,* which he had created from the Hindu name for crocodile, *Ghariyal.* The printer mistook the scientist's "r" for a "v," and the generic name appeared as *Gavialis.* It has been retained, and the creature's full name is *Gavialis gangeticus.*

The crocodilians are all amphibious—interesting examples of reptiles that returned to the water. What all have in common are a long snout, leathery skin strengthened on the back by a series of close-set bony plates, which underlie many or all of the enlarged scales, and a long, laterally compressed tail that acts like a paddle to propel them through the water.

In general, alligators have broad, rounded snouts, and crocodiles more narrow, pointed snouts. Caymans look much like their near relatives, the alligators. That's the first rough guide to determining which is a crocodile and which is an alligator. You can also spot a crocodile by observing what Lewis Carroll called its "gently smiling jaws." Do you remember the verse from *Alice in Wonderland?*

> How doth the little crocodile
>   Improve his shining tail,
> And pour the waters of the Nile
>   On every golden scale!
> How cheerfully he seems to grin,
>   How neatly spreads his claws,
> And welcomes little fishes in
>   With gently smiling jaws!

A crocodile with its mouth closed exposes the large fourth tooth on either side of the lower jaw; these teeth fit into notches on the outside of the upper jaw. But in an alligator, the fourth teeth are concealed in a matching socket in the overhanging upper jaw. The crocodile thus wears a deceptive grin. Gavials are distinctive with their slim, elongated heads. The gavial's extremely long, slender snout resembles the handle of a ladle or a frying pan stuck on a crocodile's nose. Its owner sweeps it sideways through the water to catch its fish dinners.

The eggs of crocodilians are our major interest here, but their other claims on our attention are strong. They have, as one scientist observed once, a prehistoric look, as though they had just waddled out of some Jurassic swamp. This is an exact remark because they are the nearest descendants of the awesome dinosaurs. They are the sole survivors of a once huge family of ancient reptiles called "archosaurs" (the name means "ruling reptiles"), from which descended the dinosaurs. As someone once put it, the ancestral crocodiles were the "cousins" of the dinosaurs. Crocodilians, the largest and most aggressive of today's reptiles, help us to conjure up in our minds the vanished Mesozoic world.

Their lineage is very ancient—from the end of Triassic times about 200 million years ago—and their beginnings were modest—from *Protosuchus,* a small carnivorous reptile about three or four feet long. By late Cretaceous times the crocodilians had evolved along the three lines we know today of crocodiles, gavials and alligators, and had reached a peak in the well-named "horror crocodile" (*Phobosuchus*), which, judging by fossils collected along the Rio Grande in Texas, grew to 40 or even 50 feet with powerful jaws in a six-foot-long skull. This savage giant preyed on the dinosaurs. Its length greatly exceeds that of its modern descendants; the largest of the living crocodiles are some American and Asian crocodiles—12 feet is a large one, but some in the past when they were less hunted grew to 20 feet or more. "Big Ben" was an estuarine crocodile shot in the Fitzroy River, Queensland, Australia, in the 1880s; he measured 23 feet, 6 inches and was exhibited in London. Another estuarine crocodile collected in Bengal in 1840 was said to have been 33 feet long.

Crocodilians continue growing as long as they live. An

American alligator is about 18 inches long at the end of its first year and 3 feet long at the end of its second year. Growth continues at about a foot a year to the sixth or seventh year when it is sexually mature and growth slows. They live for probably about 50 years. (One American alligator lived in a zoo for 56 years.) Exaggerated claims are sometimes made for crocodiles; some years ago photographs were published of a crocodile said to have been "old" when Shakespeare was alive!

In evolving from their modest ancestor *Protosuchus*, crocodilians acquired several features that helped their survival. One was a four-chambered heart like that of mammals and more efficient than the three-chambered hearts of other reptiles, which are liable to mix freshly oxygenated blood with unoxygenated blood. Crocodilians also have the best lungs and the best-developed brain among reptiles. They have a muscular partition separating the heart, lungs and liver from the abdomen, which anticipates the diaphragm of mammals. They acquired, too, elevated eyes and nostrils, which project just above the water and are barely discernible while the rest of the body is submerged. I've watched large estuarine crocodiles gliding submerged with scarcely a ripple, the snout and eyes looking like three floating walnuts or pieces of debris, while the half-hidden body resembled a half-floating brown tree branch—a stealthy "animal submarine." Don't think of crocodilians as sluggish. In zoos in colder climates they may be, but, since they take their blood temperatures from their environment, in the tropics they are lightning swift, lashing small animals and birds into the water with their powerful tails or rearing up out of the shallows to claw down larger creatures and drag them into deep water to drown. A large crocodilian can inflict terrible bites on large animals. With its teeth embedded and clamped tight, it rotates or spins back and forth on the axis of its hind legs to tear out great mouthfuls. The early Spanish explorers in Florida weren't exaggerating when they sent back reports of "terrible lizards." (From a corruption of the Spanish *el lagarto*—"the lizard"— has come our word "alligator.") The crocodilians have some of the best teeth in the reptile world; they're among the few reptiles whose teeth are firmly set in their jaws.

In the course of evolution the crocodilian made another important advance on their ancestors, though resembling them closely in outward appearance: The crocodilians acquired a false palate. With the help of this bony passage and a flap at the entrance to its throat, the modern crocodilian can breathe and swallow when submerged, as long as the tip of its nose is above the water. Flaps, too, seal the ears when crocodilians dive, and they have, moreover, transparent additional eyelids, like those of birds, which help protect the eyes from underwater damage should they brush against stones or sticks.

In mating, crocodilians are mainly but not exclusively nocturnal. Males attract females—and deter rivals—by loud bellowings. Some, such as the American alligator, exude a musky odor from glands on the throat and at the sides of the cloaca. The crocodilians are among the few reptiles with well-developed voices—so well-developed that one observer has likened the crocodile's declaration of love and challenge to rivals as being like distant thunder!

Look now in the shallows of the pool. By the moon's light we can see the large bull American alligator as he halts and lifts his massive head. Droplets of water, gleaming in the moonlight, run down his neck. Slowly he fills his lungs, and equally slowly he starts to roar. As the sound rises to a great crescendo, droplets of water jump from his scaly back, propelled by the vibrations of his quivering body. The great eerie roar, somewhere between that of a lion and the bellow of a bull, but more primitive than either, assaults your senses. If you were a mile away you'd hear it; near at hand you feel the ground move. It is easy to think the great dinosaurs are not dead but still with us.

The courtship, when a female has been attracted, is wild and terrifying. The two circle each other in the shallows, sending the water flying and the fish skittering away in terror. Mating, when it takes place, looks to our eyes more like a fight than a courtship. The male throws the female onto her back and unites with her. Like other reptiles (with the exception of the tuatara), the male crocodilian has a glove-finger type of grooved penis.

All over the world the numbers of crocodilians are dropping so rapidly that many experts fear they must eventually become extinct. In recent years crocodiles have gone from the lower Nile and from Egypt. Palestine and Mauritius lost their last crocodiles about 20 years ago. Their numbers have shrunk rapidly in East Africa, they're becoming fewer in India, and in American waters caymans and alligators are declining. A postwar fashion for alligator- and crocodile-leather shoes, wallets, handbags and luggage has led to wholesale hunting of crocodiles throughout the world. Most of the large estuarine crocodiles in Australia's northern waters have been wiped out. Not long ago an Australian crocodile hunter could expect to get from $60 to $90 for the belly skin of a four-foot crocodile. This is the only part of the animal that has commercial value. In 1949 one Australian shooter grossed $6000 from 924 crocodiles in three months. In recent years shooting in various countries, including Australia, has been either banned or considerably curtailed.

The wholesale slaughter of crocodilians for their valuable skins is not the whole cause of their decline; it has merely accelerated to a dangerous degree a decline that has been occurring for centuries. Man's need for land is the biggest—and most efficient—killer of wildlife. Animals must have living space to survive. Settlement destroys the crocodile's breeding grounds by clearing banks, draining swamps and polluting streams.

So serious is the worldwide threat that not long ago famous zoologists recommended to the International Union for the Conservation of Nature and Natural Resources that crocodilians should be protected during the breeding season.

More recently, in 1971, 17 of the world's leading experts met at the Bronx Zoo under the auspices of the New York Zoological Society to discuss ways and means of preserving the gravely threatened crocodilians. Among those who attended were Dr. Angus Bellairs and Dr. H. B. Cott of the United Kingdom, Dr. Robert Chadbrick of the United States, Mr. Ray Honneger of Zurich Zoo, Switzerland, Mr. Max Brown of Papua-New Guinea, and Dr. Robert Bustard of Australia.

You may ask, "Why all this concern for this dangerous man-eating creature?" Large alligators and crocodiles are not normally man-eaters. Most are more afraid of us than we are of them. The main diet of normally inclined crocodilians is fish, frogs, crabs, birds and small mammals. But if normal food becomes scarce in their habitat, then large crocodiles may turn to taking people. (Crocodiles in the East are believed to kill far more people than all the venomous snakes.) It has also been suggested that the man-eating crocodile is a rarity, like the old tiger or lion who is too weak to run down its usual game. Another theory is that the killers are crocodiles that have lost their fear of man. Be that as it may, the killing of a large man-eating crocodile is one matter; the wholesale destruction of all species of crocodilians is quite another thing. There are, moreover, paramount reasons why the crocodilian is entitled to some protection. The first is that it exists, and if we destroy it we can never replace it. Crocodilians are of tremendous biological interest. They are, as we've seen, the sole survivors of the archosaurs. They differ from all other reptiles in having a four-chambered heart and a muscular partition separating the heart, lungs and liver from the abdomen. This—and our reverence for all life—entitles them to our consideration.

The second important reason for preserving crocodilians is the ecological one. When we wipe out any creature—large or small—we cannot know what the long-range effects may be. If, in Donne's famous phrase, no man is an island, it is equally true of all creatures. Nature is a whole. Throughout nature are intricate and complex food chains and associations which can be compared to a delicate machine with many cogs. Remove or change one of the cogs and you either stop or seriously alter the operation of the machine. The balance of nature is intricate; already the citizens of some American states have discovered there were no gains and only probably losses in exterminating alligators. For instance, some citizens in Florida rejoiced when they had destroyed most of their alligators. But they gained instead a vastly increased number of poisonous snakes. What had happened was very simple: Fewer alligators had meant more frogs, small animals and birds, and these provided the food for a "population explosion" of poisonous snakes. In Louisiana, fewer alligators allowed more muskrats to breed, their burrows causing greater damage to the banks of the Mississippi. Nor was this all. The destruction of alligators in the Mississippi basin was once eagerly sought by cattle ranchers because alligators took a few calves. But the virtual extermination of the alligators brought not economic gain but probable loss; the alligator "wallows" had slowed up the summer flow of water in the streams of the Mississippi basin. With the extermination of alligators, some streams ran dry for the first time in recorded history. Some ranchers today are reported to be putting alligators back into their streams!

In Africa, too, fewer crocodiles have meant more vermin, including snakes.

The third important reason for conserving the crocodilians is that the belly skins of large crocodiles are valuable—so much so that some countries, including Australia, have started experimental crocodile farms. And the white meat from crocodiles has long been an esteemed item in native diets.

The crocodilians, as we have seen, take more care of their eggs than do turtles. They build nests where spontaneous combustion helps incubation, or they lay their eggs in sun-warmed sand. Some crocodilians also guard the eggs, and in a few species maternal care is extended to the youngsters until they are two or three months old.

25

THE SOLITARY TUATARA

T HE TEN soft white eggs, each about an inch long, that I looked at in the glass case in a New Zealand museum were not remarkable in themselves; with their semicalcareous shells they closely resembled the eggs of many other reptiles. But I lingered there for some minutes looking at (as I told myself) "the oldest land eggs in the world." There was some license in those words, but, nonetheless, if I could have gone back in time 200 million years to the Triassic period, I should have seen clutches of eggs just like these and, moreover, laid by much the same kind of reptile.

The "oldest eggs in the world" were those of the handsome, lizardlike tuatara—the world's only genuine "prehistoric monster," because this reptile, *Sphenodon punctatus*, growing as long as three feet, is the sole survivor of the Age of Reptiles and has come down almost unchanged from the Jurassic period. Fossils similar to the three-foot-long *Sphenodon* have been found in rocks from far back in the Age of Reptiles. The tuatara's ancestors were on earth before the dinosaurs;

they saw the arrival of *Tyrannosaurus,* the great flesh-eater, and *Brachiosaurus,* that 50-ton vegetarian that could see over a three-story building; and they saw the once ruling reptiles give way to the more efficient mammals.

Although it looks much like a stocky, large-headed lizard, the tuatara is more primitive in some characteristics than any lizard and is, indeed, a reptilian "missing link." It may be, as Dr. A. S. Romer suggests, the survivor of an ancient group from which the dinosaurs, crocodilians and lizards have evolved. For instance, in the tuatara's temple region are two perforations in the bones roofing the skull. Lizards have lost one of the temple openings, but both were retained in the skulls of ruling reptiles. An indication of the creature's ancient lineage is that unlike other living reptiles it has no penis.

Dr. Archie Carr has called the tuatara the loneliest reptile in the world, and with good reason; it is the sole survivor of the ancient and prolific order Rhynchocephalia, and thus, unlike other living reptiles, the tuatara has an order all to itself.

"Rhynchocephalia" means "beak-headed"—in the tuatara the front of the skull forms a kind of overhanging beak that is set with teeth. Some rhynchocephalians were much larger than the tuatara, and some had long, tooth-filled snouts.

Rhynchocephalians were once widespread but became extinct everywhere in the world except in New Zealand where a geological accident helped to preserve them. New Zealand became separated from other land bodies about 80 million years ago. As a consequence, the more efficient mammals, particularly the flesh-eating ones, did not reach there. (New Zealand has no native mammals other than some bats and, of course, sea mammals.) Elsewhere throughout the world, the more "progressive" types of animals helped to bring about the extinction of the other species of Rhynchocephalia.

Until about 100 years ago the tuatara was found all over

the North Island of New Zealand; it survives today only on some 20 rocky islets. European settlement and the introduction of predators, such as stoats (short-tailed weasels or ermines), foxes, feral cats, and dogs, led to its rapid extinction on the North Island. Only on the barren islets, which no one wanted to settle, was this archaic creature left in peace. Some 10,000 tuataras are legally protected today. Their future will depend on their island sanctuaries being left undisturbed.

The short-legged adult tuatara is a picturesque creature with a body color of greenish-brown, spattered with dark-green and sulphur-yellow spots and splotches. A large crest of brilliant white spines grows along the neck and back and has earned for the animal both its scientific name *punctatus* and its Maori name of "tuatara," meaning "peaks on the back." The ruffled white crest gives the tuatara a formidable look, but actually the spines are soft and flexible. The tuatara's big, heavy head, has two large, brilliant eyes. Even more singular is its third or median eye on the top of the head and covered with skin. Common in various lizards and thought to be the remnant of a second pair of eyes, the organ has a lens and retina but apparently does not function.

The tuatara is a somewhat sleepy creature—its metabolism is presumably even lower than that of most reptiles. Even when you pick up a sleeping one it awakens only slowly. However, it is capable of a short, wobbling burst of speed when frightened or pursuing its insect prey. It hunts mostly at night and feeds somewhat austerely on beetles, grasshoppers, spiders, flies and other insects. Because of its low metabolism, it doesn't need much food; some in captivity seem to be quite happy with a couple of snails or a few pieces of raw meat a day during the summer. (The appetite is keenest in summer; by May—the down-under autumn—the tuatara has usually lost interest in food and now retires to hibernate until spring.) One in the Phoenix Park Zoological Gardens, Dublin, Ireland, earlier in this century lived for 28 years on six earthworms a week and an occasional snack of hawthorn leaves, but in its natural habitat its diet is wholly carnivorous.

Only the mating season seems to stir the tuatara into much activity, at which time it will fight occasionally with another male and bite savagely. It is not only strong and handsome but usually silent—its only call being an infrequent grunt or croak. It takes an easy course with its housing and often lives in the nesting burrows of shearwaters or petrels (*Puffinus carneipes* and *P. bulleri*), those ocean-ranging birds that breed in the tuatara's islands during summer. The tuatara thus has the burrow to itself during the winter, and, because of their different habits, share it fairly amiably with the rightful occupants during the summer. When the shearwaters have nestled in on their eggs for the night, the resplendent, ruffed tuataras emerge from the burrows to sup on insects. Petrels, however, sometimes eat the tuatara's eggs. Sometimes, when it can't find a burrow, or when its shearwater roommate is too bad-tempered, the tuatara digs a rather shallow hole in sand or loose shingle.

The female lays between 8 and 15 soft, white eggs, each about one inch long, in a pit she digs and then covers. They are laid in November (there, the late spring), are left unattended and do not hatch until 15 months later during the following summer. The development of the embryo is rapid for about four months and then almost stops until the following spring. This "hibernation" within the egg is rare and is found only among a few reptiles.

At first, the young tuataras rather resemble tortoises; only much later do the more lizardlike characteristics appear. Their chocolate-brown bodies are marked with longitudinal and transverse bands of grey and white—a protective pattern they lose as they merge into the splendid adult form. Unlike the adults, young tuataras are very active. They grow quickly at first but much more slowly in succeeding years, and take about 20 years to reach sexual maturity.

Perhaps these lone survivors from the Age of Reptiles have some secret of longevity. Some in captivity have lived for over 50 years—a century is probable, in view of the long time they take to reach sexual maturity. The Maoris, in fact, claim that they live for hundreds of years.

The eggs may be the "oldest" laid by any living creature, but in other respects the reproduction habits of tuataras do not differ greatly from other reptiles—that is, they lay their eggs and leave the young to take their chances. And presumably only isolation in New Zealand without competition from more efficient creatures has enabled them to survive almost unchanged for 135 million years.

26

LIZARDS

THE SOLITARY TUATARA, the crocodilians and the turtles are "living fossils"—the remnants of a once great army of ruling reptiles that dominated every corner of the land, sea and air in Mesozoic times. The other order of living reptiles, the Squamata (the name means "scaly-skinned"), which embraces lizards and snakes, is much more modern. Lizards don't appear before the Jurassic period, which began 180 million years ago when the first birds were appearing and reptiles ruled, and snakes do not make their appearance until towards the end of the lower Cretaceous period, about 100 million years ago. The Cretaceous was the last of the three periods of the Age of Reptiles.

Lizards are regarded by biologists as comparatively "unprogressive" because they retain primitive features, including their sprawling type of walking and running. In spite of these drawbacks, they have been comparatively successful. They are the most abundant of living reptiles, with over 30,000 species in the tropics and in temperate regions, where they resist the winter by hibernation. Snakes, which evolved from lizards, are the most progressive—and successful—of living reptiles in ways we'll examine in the next chapter. (It helps to understand snakes if you think of them as rather specialized lizards.)

Most lizards lay eggs. The shells of the eggs are either soft and leathery or hard and calcareous (limy). Just as their reptilian ancestors did, lizards bury their eggs so that the warmth of the soil will provide the heat for incubation. Some lizards, such as the great teju (*Tupinambis teguixin*) of South America and the Nile monitor (*Varanus niloticus*), often lay their eggs in termite nests, where the temperature is usually higher than that of the surrounding soil.

With most lizards, motherhood ends with the laying of eggs. A few, however, are better mothers. For instance, some skinks (*Eumeces* spp.) of America and elsewhere coil around their eggs during incubation and guard them from intruders. They move the eggs about constantly and leave them from time to time to bask in the sun, and then return to transmit the sun-absorbed warmth to their eggs. (A familiar skink who does this is the blue-tailed skink.) One American skink even approaches the level of some of the crocodilians in its care for its young; the female helps the babies to bite their way out of the eggs, guards them while they feed and even looks after their hygiene by cleansing their cloacas with her tongue.

The maternal instinct is so strong in female skinks that a female gathers to her own clutch any egg left unguarded by another female. And a female skink knows a skink egg from another kind even when her eyes are covered—as experimenters have discovered.

Other American lizards that coil around their eggs are the so-called glass snakes (*Ophisaurus* spp.), which are really legless lizards. (British relatives of the North American glass snakes are known as slowworms, [*Anguis* spp.].) Glass snakes are unfortunately sometimes killed by people who mistake these harmless creatures for snakes, but they are easily identified because, unlike snakes, they have movable eyelids and external ear openings. A snake's eye is protected by a single transparent scale. These eye discs are shed each time the snake sloughs its skin. (The staring eyes of snakes have given rise to the erroneous notion that snakes hypnotize their prey.)

A lizard advance on turtles and crocodilians is that some bear living young—with all the advantages that this usually confers. The young lizards developing in their mother's body are much safer than, for example, unguarded turtle eggs. If you live in southwestern America you will probably be familiar with the so-called horned toad (*Phrynosoma cornutum*), which is no more a toad than you are but is, of course, a lizard. Admittedly, with its broad, flat body, rough skin and squat head, it is not unlike a toad in general appearance. It lives on ants. The female in the breeding season digs a hole up to a foot deep, lays her leathery-shelled eggs and covers them with earth. This is typical lizard behavior, but in New Mexico there is a closely related species of horned toad that is ovoviviparous —the young are born alive. The New Mexican horned toads have no glands for forming shells; the eggs are incubated in the mother's body and the young born alive each in a thin, transparent sac. The moment they emerge they begin hunting ants. You can think of them being born alive and packaged in something that looks like cellophane!

Some skinks, too, bear living young. Some have a type of primitive placenta that closely resembles that of some of the marsupials. (Marsupials, you will remember, bear their young alive, but in an immature state—so much so that a baby kangaroo may be only one and a quarter inches long. The further development of the kangaroo takes place in the pouch to which the tiny gelatinous-looking baby makes its way.) The lizard embryo derives some nourishment from its mother, but most comes from the large yolk sac.

Lizard eggs, whether developed in the mother's body or hatched outside, must be fertilized internally, and this involves cooperation between male and female. Reptilian courtship is at its gayest among lizards. Some male lizards have breeding territories, much as male birds do, and they defend them with displays that play much the same part as birdsong does for the many birds—that is, the displays deter other males and attract females. Sometimes fights ensue between male lizards, but there is usually more bluff than bite. Some chameleons stage pushing matches, usually without biting, and display as brilliantly as Siamese fighting fish. The males' color changes, and females are attracted to those with the brightest dewlaps. Other male lizards raise their crests, display colored throat fans and arch their backs. Still others court with stiff-legged dances. (Many but not all lizards have color vision.) One of the most spectacular displays is that of the frilled or blanket lizard (*Chlamydosaurus kingii*) of Australia, which spreads a frill resembling the ruff of an Elizabethan courtier. A lizard two and a half feet long may have a frill one foot in diameter, gorgeously splotched with pink, black, brown and white against a black chest and throat. When displaying—or at bay against intruders—the lizard shows a pink-lined mouth. Sometimes it rises on its hind legs and rocks from side to side. It accompanies this display, which looks impressively fierce to an intruding dog, with loud hisses. The brilliant frill is thought to serve other purposes in addition to intimidating male arrivals and intruders and impressing females; it is believed to help the frilled lizard control its body temperature by shedding excess heat.

Among other lizards the breeding territories are held not by males but by females. The female must be wooed on her own piece of real estate. Among some species of fence lizards, the female landowners attack all males who approach for most of the year; only when she is ready to mate does she permit a male on her property.

At least two species of lizards have no males, and their young are almost certainly produced by *parthenogenesis*—that is, the embryos develop without fertilization of the eggs. This is the normal practice among aphids and rotifers, which also have no males. Parthenogenesis also occurs among various insects—the honeybee is a famous example; the queen's fertilized eggs produce the workers, and the unfertilized eggs the males or drones. Parthenogenesis is suspected in certain geckoes.

Among lizards the courtship is visual at first; sound plays little part because most lack true voices. (Geckoes, some skinks and legless lizards are the only lizards with true voices.) Then, as the pair come together, odor takes over.

Male lizards—and male snakes, also—have *two* glove-finger penises called *hemipenes*. When the female has been wooed, the male seizes her by the neck and twists his tail so as to insert one of the hemipenes into her cloaca.

Lizards were comparatively late arrivals. None are known as fossils from earlier than the Jurassic period, about 150 million years ago. While they retained the comparatively awkward sprawling gait of their ancestors, they acquired lighter limbs in the course of evolution, so that some modern lizards have a reasonable rate of speed. Most have two pairs of legs. If there was a lizard Olympics, the 100-yard dash would be won by the aptly named racehorse goanna (also called Gould's monitor, *Varanus gouldii*), which is credited with reaching 20 miles per hour on the sandy plains of Australia's inland. The bewildering speed of this dull-brown and yellow monitor helps it considerably to run down its prey of other reptiles and small mammals of the open plains. (Monitors are sometimes called goannas in Australia—a corruption of "iguana," to which they are in no way related.) Other speedsters are the zebra-tailed and collared lizards of North America, which have been timed at over 16 miles per hour.

These lizards rise up on their hind legs when running. Another in Australia, called the "bicycle lizard" (*Amphibolurus rufescens*), gets its name because it runs on its hind legs when alarmed and looks as though it is pedaling frantically.

Other lizards are speedy climbers. The lace monitor or tree-climbing goanna of Australia (*Varanus varius*) scampers up the straightest and smoothest eucalypt trunks with ease. Flushed on the ground, this monitor, which grows to seven feet, races for the nearest tree. Sometimes in panic it claws its way up the nearest person, whom it presumably mistakes for a tree stump. (Not long ago an Aboriginal woman in central Australia had several stitches taken in her body and head after being climbed by a frightened lace monitor.) The lace monitor gets its name from its markings—usually black with irregular bands of yellow dots.

Monitors have long mobile necks and tongues that are deeply forked like those of snakes. It's probably this snakelike tongue that has led to the erroneous belief that the bites are poisonous and that they never really heal.

Only two lizards are known to be poisonous—the Gila monster (*Heloderma suspectum*) of the southwestern United States and extending into Mexico, and its close Mexican relative, the beaded lizard (*H. horridum*). Both are well named. These stumpy nocturnal lizards eat small mammals, birds and eggs. The toxin is injected into their victims by grooves in their teeth, which, unlike those of venomous snakes, are in the bottom jaw. The Gila monster is, of course, named after the Gila River in Arizona. Both lizards lay up to 15 yellowish-gray eggs toward the end of July, which hatch about a month later.

The largest monitors are those of Asia and Australia, growing in some species to 7 feet or even 12 feet. They are fierce predators of small warm-blooded animals, although they are also scavengers and feed on carrion. They are powerful creatures—a swipe of a large monitor's tail has smashed a dog's leg and frequently knocked men off their feet. One 6-foot Australian monitor was observed to run down and kill a partly grown kangaroo.

The largest of all living lizards are the monitor lizards of Komodo Island and nearby islands in Indonesia, which attain a length of about 12 feet and weigh up to 300 pounds. Sometimes called "Komodo dragons," these monsters are fierce and unfastidious carnivores, feeding on living animals, such as small deer and pigs, and on carrion. One knocked over on the ground a large macaque monkey, crocodile-style, and swallowed it whole. Another brought down a small deer and swallowed it whole. Yet another devoured the whole hindquarter of a dead goat. Komodo dragons, incidentally, are the world's most expensive reptiles; not long ago a Djakarta breeder was asking $3000 for a single one.

The second-largest monitor is also Asian—the two-banded monitor or kabaragoya (*Varanus salvator*) of Southeast Asia and Malay Island to the Philippines—which attains a length of ten feet.

(By the way, there's some uncertainty about how the name "monitor" came to be given to these big lizards. One claim is that the word, which of course means "watcher," was given to the Nile member of this large worldwide family in Roman times because it hissed whenever crocodiles approached and thus gave warning.)

Like other reptiles, lizards were much bigger in prehistoric times. One lizard from the Pleistocene epoch of Asia, about one million years ago, was about 24 feet long and thus as large as many dinosaurs. Largest of all ancient lizards were the mosasaurs, giant marine lizards of the Cretaceous period. These lizards must have been very successful in their return to the sea, because their fossils have been found in almost every corner of the earth. The name "mosasaur" is derived from the discovery of the first specimen on the banks of the Meuse River (Latin *Mosa*) in Holland. Mosasaurs grew to 30 or 40 feet in length, and probably preyed on fish and other reptiles. About half their length was in the long, flattened tail, rather like that of a modern crocodile.

Other small prehistoric lizards also returned to the sea, and were successful for a time. Today the only marine lizard lives in the Galapagos Islands and feeds on algae. It swims well with its long, flattened tail but is not completely adapted to a marine existence; its feet are only partly webbed.

Some other living lizards lead a semiaquatic existence—notably some of the large monitors of Australia and the tree iguanas of tropical America. Some skinks, too, spend much of their time in the water, hunting for food.

Although relatively primitive, lizards make up the largest group of living reptiles, with an enormous range of structural types, some 300 genera and 3000 species. They have colonized most of the tropical and temperate world. In deserts they defeat dehydration by burrowing and by hunting at night; some desert skinks capture up to 600 insects in 24 hours and acquire water from them as well as food.

A useful adaptation acquired by many smaller lizards and geckoes is the ability to shed the tail, which continues to wriggle energetically for some minutes. This often allows the lizard to escape while its enemy's attention is captured by the wild wriggling of the tail. Later a new tail grows; it is sometimes slightly shorter than the old tail and differs a little in color. Once in a great while something goes wrong and a gecko may grow two tails; in the Philippines a two-tailed gecko is regarded as a bringer of good fortune, and house-holders like to have one around.

Some modern lizards, too, highlight the fact that birds developed from lizards. The flying dragon (genus *Draco*) of Southeast Asia has large folds of membrane supported on five or six pairs of ribs, which it extends from its thin body. In full "flight" between tree trunks it looks not unlike a delta-shaped aircraft. With its "wings" outspread, a flying lizard can glide 50 or 60 feet from trunk to trunk.

Lizards, like other reptiles, are "cold-blooded," but they are not as much at the mercy of the environment as the adjective suggests. They show great skill in choosing their environments so as to ensure suitable body temperatures. Scientists have implanted tiny thermistors in the body of an Australian lace monitor and connected them to a miniature radio transmitter. By means of a direction-finding receiver, an Australian zoologist, Dr. Richard Barwick, was able to plot both the track and temperature of the monitor for many days. Moving between sunlight and shade, the reptile was able to maintain a body temperature about five degrees below our own throughout the day.

Lizards are economically important because almost all of them feed on insects. This is true even of some of the larger ones, such as the monitors of Africa, Asia, Indo-Malaya and Australia. And all, too, are valuable disposers of carrion. And all are important from the ecological point of view.

27

SNAKES

JUST AS LIZARDS DO, male snakes also compete for females. Males indulge in spectacular and ritualistic wrestling. Two males will pursue each other over rocks and through undergrowth and even through streams. When one catches the other, they intertwine, wrestling and writhing so vigorously that the grass is flattened. Eric Worrell told me that at his Australian Reptile Park near Sydney these wrestling displays, which continue daily for about a month, are so frenzied that it is necessary to repair the landscaping of the snake pits!

Diamondback rattlers (*Crotalus adamanteus*), of North America and the southern United States, engage in knightly contests in which the snakes entwine and try to climb above each other. I use the word "knightly" because in these combats the diamondbacks (and other poisonous snakes) do not use their poisonous fangs on each other.

The energetic wrestling contests of male snakes presumably determine the fittest male and are of survival value to the species. The snake who loses slinks away.

219

While the suitors have been fighting, the female may have put on a milder display of her own of sinuous writhings. This may attract the male, but generally among snakes sight is secondary to smell because they are relatively shortsighted. The male snake tracks his bride—and also his prey—by smelling with his nose and with his wonderfully sensitive forked tongue, which is both a nose and a hand. The flickering tongue scoops microscopic particles of scent from the ground and air, and the tips convey them to two tiny sensory pits (the vomeronasal organ) in the roof of the mouth. Both male and female caress each other; the male insinuates his body back and forth along the full length of his bride and rubs his chin along her back. During courtship some pythons rub their thornlike vestigial limbs against each other. Copulation is accompanied by vigorous tail twitching as the male inserts one of his two hemipenes.

Like some turtles, some female snakes may produce fertile eggs although they haven't mated for several years. And like lizards, most snakes reproduce by laying eggs. Those in the United States that lay eggs include the poisonous coral snakes and the nonpoisonous whip snakes, racers, rat snakes, king snakes and bull snakes. Those that lay eggs usually worm their bodies into loose leafmold, a hollow in decomposing wood, or beneath a suitable clod of earth—places where decaying vegetable matter will generate heat to help incubate the eggs. Most snake eggs are elongate in shape and white when first laid, with a leathery or parchmentlike shell. These shells are flexible and consist of thin laminated layers and elastic fibers with a fine limy coating.

Maternal duties usually end with the laying of the eggs. A few snakes, however, do a little more for their eggs; just as the skinks do, they brood their clutches. All the pythons, for instance, brood their eggs. Like skinks, they constantly shuffle their eggs about (which raises the temperature of the eggs some three to five degrees above the environment) and the female python leaves her eggs on sunny mornings and sunbathes for an hour or two. Before she leaves the eggs she covers them. Having absorbed heat from the sun, she returns to twine once more around her eggs. Pythons guard their eggs devotedly, and if you approach a brooding mother too closely she will strike out fiercely and sometimes bite you. Her fangs aren't venomous but can inflict a nasty wound. (Like the closely related boas, pythons kill their prey by constriction and swallow it whole without chewing.)

Most snake species lay from 8 to 15 eggs, but the pythons go in for large clutches, and the bigger the snake the more eggs she lays. For instance, a female African rock python (*Python sebae*) may lay about 20 eggs, each larger than a goose's egg, when she is 14 feet long and up to 100 when she is 8 feet longer. An Indian rock python (*Python molurus*) laid over 100 eggs and guarded them 11 weeks while they incubated. For some days the newly hatched young crawled back at night into the egg shells, which the mother continued to guard. (The Indian rock python is the one most used in sideshows for so-called snake-charming acts.) Pythons live in Africa, Asia, Australia and the Philippines.

North American snakes that brood their eggs include several species of the genus *Pituophis*, called bull snakes. These snakes, growing to nine feet, lay a dozen or so creamy-white eggs and curl around them for up to three months.

Another good snake mother is the king cobra or hamadryad (*Naja hannah*) of Asia, growing to 18 feet, which builds a nest of grass and leaves in which to lay 20 to 40 eggs. Both parents usually remain together until the eggs hatch. The female coils in the nest chamber; the male joins her there or waits nearby.

Other snakes that brood their eggs include certain vipers, cobras and colubrids. (These last belong to an immense family, Colubridae, which includes most of the snakes of the world. Most of these snakes, which include water snakes, grass snakes and tree snakes, are harmless to man.) In some instances the body temperature of the mother snake is said to rise by a few degrees, but this is probably of only marginal help in incubation; the major advantage of brooding is the safeguarding of the eggs.

In temperate climates egg-laying snakes and lizards usually lay their clutches in early summer. The length of incubation ranges widely from a few days in some species to as many as 90 in others. The average is about 56 days. In any single species the time for hatching depends on outside temperatures and is therefore necessarily variable.

Both baby snakes and lizards have tiny and temporary egg teeth with which they cut their way out of the leathery shell.

Some snakes don't lay eggs but instead give birth to living young. The live-bearers—about a quarter of the 2700 snake species—include the boas of tropical America. They're such near relatives of the pythons that zoologists put them all in the one family, Boidae, and distinguish pythons from boas by a bone above the eye of the former, which is lacking in boas.

Both pythons and boas have vestigial hind legs looking rather like a thornlike spur on each side of the vent. These vestiges of hind legs—and their well-developed lungs—single out the Boidae as the most primitive of living snakes. Most snakes that are more advanced have lost all external traces of hind limbs and have the right lung elongated to fit the slim body and the left lung much smaller or even absent.

In general terms, there would appear to be advantages in bearing young alive, yet egg-laying pythons appear to be just as successful in their habitats as live-bearing boas are in theirs. Sometimes the reason why a reptile should bear live young is obvious. For instance, the European adder (*Vipera berus*), the common lizard (*Lacerta vivipara*) of Europe and the slowworm are three reptiles who make their homes in the colder parts of the Northern Hemisphere. The European adder and the common lizard (also called the "viviparous lizard") even venture into the Arctic Circle. For much of the year the ground in the northern parts of countries such as Norway, Sweden, Denmark and Scotland is much too cold to provide much warmth for eggs buried in the ground. All three are therefore live-bearers. However, in the more temperate parts of Europe, the common lizard lays eggs.

Snakes in North America that bear live young include the poisonous rattlesnakes, copperheads and cottonmouths and the nonpoisonous water snakes and garter snakes. They give birth to miniature adults who can fend for themselves from the moment they break the transparent envelopes in which they come into the world. One of the most fertile of snake mothers was a garter snake who gave birth to 78 in one litter!

Most sea snakes are live-bearers. These are the advanced sea snakes that biologists call the true sea snakes. They are, of course, land animals that returned to sea and made a successful adaptation. For instance, they have flattened bodies, large, powerful paddlelike tails and are so skilled at swimming that they are helpless on land; they slither about futilely. The sea snakes that lay eggs on land are called the false sea snakes. They are not so adapted for swimming and are quite agile on land.

There is, of course, one obvious advantage for a sea snake in bearing live young—it freed them from the land.

Like their lizard counterparts, most of the snakes that produce live young are ovoviviparous, and the eggs don't develop the tough outer skin of eggs that are hatched outside the body. The eggs are retained in the oviducts where the embryos develop fully. In some instances the embryo derives all its nourishment from the yolk sac; in others there is a primitive placenta from which the embryos receive some nutriment.

Because all snakes are wary, it is rare to see live-bearing snakes giving birth to young in the wilds. However, they have often been observed in zoos. The following is a graphic account by Mr. Eric Worrell in *Reptiles of Australia* of an Australian mainland tiger snake (*Notechis scutatus scutatus*) giving birth to live young.

> The female chooses a shaded position, almost obscured by grass, among gnarled tree-roots. She relaxes full length across some of the smaller roots. A number of newly born young crawl into the grass, one across her body. She elevates her tail and almost effortlessly a transparent sac, in which the banded body of a fully formed snake is visible, emerges from the greatly distended anus. The hindermost portion of her body contracts as the sac is expelled. The baby snake once clear of the mother fights and wriggles against the thin membrane until it frees itself, then, still wet and slimy, spreads its hood to dry and assumes a mock striking stance, darting at imaginary enemies. There is no regular period between births, each individual snake taking only a few seconds, while the swelling of the next can be seen working down the body of the female. These young snakes are able to care for themselves from the beginning. Like other poisonous snakes, they are born or hatched with fangs and venom glands ready for use.

Parental care in most snakes doesn't extend beyond birth or hatching. In a certain natural history museum there is a touching exhibit of a mother venomous snake basking on a rock, surrounded by 15 of her progeny. Alas, there is no such maternal aftercare; young snakes would not sleep in the sun close to a large adult that would make a meal of them if it was hungry. Young snakes are, moreover, well able to fend for themselves.

There is a myth that some mother snakes swallow their young when danger threatens. This has probably come about because when a female live-bearing snake is killed, live young are sometimes found inside her.

Growth among snakes is rapid in early life. They may double their hatched length in their first year and treble it in their second year. Thereafter growth slows, but never ceases entirely. The longer they live, the bigger and longer they grow.

Snakes, as I said earlier, are comparative newcomers in the evolutionary story and are best thought of as much-modified lizards. They do not appear until the late Cretaceous, when the Age of Reptiles was passing. The fossil record is scanty because their frail heads were rarely preserved.

No one knows how snakes came to lose their limbs, but the generally held theory is that the ancestral snakes lost their limbs when they became burrowers in the earth. It is thought, too, that snakes had a second evolutionary phase that gave rise to the great variety of snakes we have today. Whatever the reasons for limblessness, snakes have acquired some compensations: Lack of limbs allowed them to grow long; some, such as the pythons and boas, acquired new ways of killing their prey—by constriction; and their new slim bodies, narrow because they lacked breastbones, enabled them to find shelter and security in narrow places where other creatures couldn't follow. The long body also enabled some to take to the trees. And though snakes aren't very fast, their rippling motion and lack of encumbrances enables them to move faster than most creatures among rocks and thick undergrowth. Have you ever tried to outrun a snake in rough country? I have—and wasn't even in the race. But most accounts of the speed of snakes are wildly exaggerated. Racers are believed to be among the fastest and most active snakes, yet they have only clocked 3.6 miles per hour in tests. Snakes look fast because they can explode into top speed instantly and can move comparatively quickly in rough ground. However, they are capable of only short bursts of speed because their lungs aren't very efficient.

Snakes have no outside limbs, but they do have internal ones—as many as 280 or even 800 or more. A curious adaptation brought about by the snake's slender shape is that the viscera are elongated to conform, and the left lung, as we noted earlier, is either smaller than the right or is absent. The supple backbone of a snake may have 140 or even over 400 vertebrae, each with a pair of ribs linked by powerful muscles to the belly scales, which grip the ground. A snake wriggles forward by drawing together and expanding alternate groups of ribs. The body undulates into S-like curves whose outer surfaces are pushing points. The snake thus goes forward in a series of wavelike ripples in much the same way as an eel swims. On a polished floor snakes thresh about purposelessly because their scales have nothing to thrust against. Some land snakes, particularly the larger ones, move forward rather like a worm—in a straight line, by expansion and contraction of the overlapping belly scales which move in wavelike ripples.

Desert snakes have another form of motion, called "sidewinding," such as that of the sidewinder rattlesnake (*Crotalus cerastes*). It corkscrews or rolls the coils of its body over the ground in a sideways motion.

A fourth form of motion is "concertinaing," in which the snake reaches out to an anchor point, fastens to it and brings the rest of the body up to this point. This motion is used mainly in trees.

Because a snake can't run down its prey, it must either ambush or stalk it. Snakes such as the pythons and boas, as we've said, wind their coils around their prey and suffocate it. The venomous snakes paralyze their prey with toxin. The acquisition of venom, which is a highly modified saliva, is one of the more recent adaptations. Snakes that lack venom have to seize their prey with their jaws and engulf it.

The snake's world is soundless. It has no outside ears but can feel vibrations through its body. They locate prey by sight or smell, but because of their shortsightedness, rely mainly on smell. I once watched an Australian tiger snake trail a marsupial rat for 50 feet by smell alone, flickering its tongue just as a dog might sniff the air. Its tongue flicked in and out of its mouth, where the tips conveyed the sensation to the two tiny chemoreceptor pits in the roof of its mouth. It found the rat, struck—and waited. The rat ran a few feet and died. (The tiger snake reputedly has, drop for drop, one of the deadliest venoms in the world. The average yield of a milked snake is 35 milligrams, enough to kill up to 80 men. A few colorless drops could also kill up to 20 horses, or 118 sheep, or 23,000 white mice. The venom is five times as toxic as that of the famed common cobra.) A brilliant adaptation in the pit vipers and some boas and pythons is a heat reactor that can register small differences of heat, even in total darkness. They are thus able to find warm-blooded prey at night.

Other remarkable adaptations allow a snake to swallow prey much greater in diameter than the snake's head. This enables a snake—which may be able to catch prey only at infrequent intervals—to take in a huge meal whenever it is fortunate enough to get one. Thus snakes, with their cold-blooded slow rate of metabolism can fast for weeks or even months; they need a comparatively small amount of food in proportion to their bulk in order to survive. A python in a zoo is reported to have gone without food for four years.

The reason a snake can swallow prey much greater in diameter than its skull is that some of the bones in the skull are movable and the lower jaws are joined by an elastic ligament. These jaws can stretch widely, and, moreover, the left and right jaws are capable of independent movements. Snakes, as we have seen, have no crushing or chewing teeth and must swallow their prey whole. While the prey is gripped by the left jaws, the right jaws can reach forward and establish a new grip. This is repeated by the left jaws, which release their grip and extend forward and take a fresh hold. The backwardly curved teeth keep a constant grip on the food. Large pythons and boas can swallow quite large animals whole. Anacondas, for instance, prey mainly on capybaras, those largest of all rodents, which are about the size of an average domestic pig.

You can dismiss as tall stories accounts of anacondas swallowing horses and cattle—or men. Anacondas—and large pythons—growing up to 30 feet long, could conceivably swallow a medium-size man, but rarely do. Ramona and Desmond Morris discuss this in their book *Snakes and Men*:

> The maximum capacity of a really big constricting snake is a prey object of about 150 lb., although even the largest snakes have been seen to have some difficulty in swallowing a meal of 100 lb. From this it would seem that a human being of moderate size might be fair game, and there is no doubt that a wild constricting snake of upwards of twelve feet could overpower the average man, unless he were in excellent physical condition and knew how to prevent the animal from obtaining a grip with its jaws. Yet few human beings are in fact attacked and eaten by giant snakes.

For an explanation of why this is so the Morrises turn to the acknowledged expert on large snakes, C. H. Pope, who points out that the answer probably lies in the size and intelligence of man. Human shoulders, Pope says, are probably too wide for boas of average size. The Morrises cite one of the few authentic accounts of the swallowing of a human being. A 14-year-old boy was missing in the East Indies (now Indonesia), and a 17-foot python was later seen with a suspicious bulge. When the python was killed, the bulge was found to be the body of the missing boy. One of the most gruesome stories tells of a man in the tropics who fell into an alcoholic slumber and awoke to find that a large python had swallowed one leg as far as it could go. The man's screams on awakening brought rescuers who slew the python—but, alas, it was too late; while the man had slept, the snake's digestive juices had been working on the leg. Only an amputation saved the man's life. Is it true? Probably it is a mixture of fact and fiction, an incident that got better with each retelling.

There's an authentic account of a large reticulated or royal python (*Python reticulatus*) of India and Malaysia swallowing a 123-pound deer. And Frank Buck, who captured animals for the world's zoos, once had a large pig given to him by grateful Indian villagers after he'd captured a man-eating tiger. He put the pig in a pen made of stakes driven into the ground, and the next morning he found he'd "captured" a large python. It had slithered easily through the stakes, swallowed the pig and had then been too large to get back out. Buck was delighted. He had wanted a large python for a zoo and had been prepared to offer many pigs to anyone who brought one to him.

A virtuoso swallower is the common egg-eating snake (*Dasypeltis scaber*) of Africa, whose staple diet is eggs. This brown snake, growing to about two and a half feet will swallow a bird's egg with a few convulsive wriggles, and then, not long afterward, regurgitate the broken and empty shell. Some of the snake's neck vertebrae run down into the gullet and are equipped with spikes. These pierce the egg just after it has been swallowed. After the contents are swallowed, the snake ejects the battered shell. The egg-eating snake is said to reject any egg which is even slightly off.

One of the bull snakes, the gopher snake, *Pituophis catenifer sayi*, has no convenient egg-breaking spines but manages quite well anyway. When the egg has been swallowed whole and is an inch or so down the gullet, the snake thumps its neck against the ground, swallows the contents of the cracked egg and discards the smashed shell.

Of the 2700 species of snakes, about 500 are venomous,

but only 200 are dangerous to man. Contrary to popular belief, snakes don't go around biting at random; they use their venom to capture their food, and know that we are too big to eat. At Eric Worrell's Australian Reptile Park at Gosford, New South Wales, guinea pigs live in a pit with four-foot tiger snakes. The guinea pigs are put there to keep the grass short, and the snakes never strike them because they know the guinea pigs are too big to be swallowed.

Snakes try to avoid us and usually do. If you see one in the country on an afternoon's walk, a dozen or even a hundred have seen you and glided silently away.

The truth is that snakes are shy. When a snake does bite a man it is either a physical or a social accident. Sometimes we tread on one before it can get out of our way, and the frightened snake strikes in defense. Significantly, most snakebites occur in early spring when temperatures are low and the reptiles, emerging from hibernation, are sluggish. What I have called a "social accident" happens because we don't understand the snake's code of signals. "Most venomous snakes are reluctant to bite but will do so if you insist," Charles Tanner, a noted snake expert of Cooktown, Queensland, Australia, told me. "They usually go through a warning routine first such as hissing, a warning gesture or a false strike. If you don't retreat or allow them to retreat and still appear to be attacking, they may bite."

Among the poisonous snakes, only the larger ones can give us a bad bite. The worst bite in the world probably belongs to a sea snake (*Enhydrina schistosa*) that inhabits the coastal waters of a huge triangle bounded by northern Australia and the western Pacific, the Persian Gulf and South Japan. Its nerve poison is twice as toxic as the tiger snake's and ten times as toxic as the common cobra's. But the snakes are shy, and bites are rare. Malayan fishermen haul them up in their nets, grab them by the tails and hurl them back. Sometimes one is bitten, yet only one in three die, and most of the others develop few or no symptoms of poisoning. "This is because the snake doesn't give them a business bite," explains Dr. Alistair Reid, a London physician who carried out a survey of these sea snakes.

Like lizards, snakes are of considerable economic importance. A few are dangerous, but they are outnumbered by the others, which feed largely on insects and rodents. A zoologist once estimated that every gopher snake was worth about $50 a year to U.S. farmers in some regions as a destroyer of rodents.

Of living snakes, the boas and the pythons are thought to most resemble their primitive ancestors. One of the largest prehistoric snakes was *Gigantophis* of the Eocene epoch of Egypt of about 60 million years ago, which is estimated to have attained 50 feet. No modern snake measures anything near as long. The reticulated or regal python is credited with a length of 32 to 33 feet. The second largest is the anaconda (*Eunectes murinus*) of South America with an authentic record of 28 feet. (There are claims of an anaconda said to have been 37 feet long. Colonel Fawcett, the British explorer who disappeared in Brazil in 1925, claimed that he measured one at 62 feet.) Other maximum lengths given by the Zoological Society of London are Indian python, 25 feet, and African rock python, 25 feet.

Like other reptiles, snakes and lizards are long-lived. The lizard record is possibly held by a slowworm that lived in the Copenhagen Zoo for 54 years. An Indian rock python lived in captivity for 31 years, and an anaconda lived for 28 years in the Washington Zoo.

To sum up, a few lizards and snakes guard their eggs and need to lay fewer eggs than those lizards and snakes that abandon them after laying. As we have seen, most lizards and snakes lay eggs where the sun's warmth will hatch them or in decaying vegetable matter which generates heat. A few brood their eggs much as birds do, and a few bear their young alive.

PART THREE

Ultimate Wonder in Birds' Eggs

28

THE FIRST BIRDS

As WE HAVE SEEN, most reptiles lay their eggs and leave their young to fend for themselves. The big leap to parental care of the young was taken by the descendants of reptiles—by birds, which have been called "feathered reptiles" and "glorified reptiles." The last remark has been credited to T. H. Huxley. Both probably offend the more sentimental bird lovers. Birds are the most beautiful, the most engaging and the most observed of all the creatures with which we share this earth. Nonetheless, the reptilian ancestry of birds is obvious even if you look no further than their scaly legs. They also have certain features in common with reptiles in their muscles and skeletons, lay similar eggs and have an egg tooth in the upper jaw just before hatching. Unlike reptiles, however, they are warm-blooded creatures like ourselves—the average body temperature of birds is 106° F., which is higher than that of mammals, including ourselves; and unlike all other creatures they have feathers.

Whereas almost all reptiles leave their young to fend for

themselves, birds care for their young. This is true of all but a few species. Linked with this care of the young is the building of nests to help keep the eggs warm and to protect the young. Moreover, the development of nesting is a logical accompaniment of flight. The young need to be cared for while they grow flight feathers and acquire the complicated skills necessary for flight.

When you think about it the step to nest building was the only one birds could have taken. Lightness is everything in a creature who has to fly. The bones of flying birds, for instance, are light and hollow. If birds had tried to bear living young, the mothers would have been disastrously handicapped in flight and easy prey for predators; as a consequence birds never developed along these lines as the mammals and some reptiles did.

Nest building is so diverse and specialized among birds and plays such an important part in our story of eggs that it will be given a later chapter on its own. The present chapter is concerned with how birds evolved from reptiles.

We do not know the full story of how reptiles developed and became birds. This is a "ladder" with many missing rungs because few complete fossilized skeletons of birds have been discovered. The bones of birds are so fragile that they are easily broken or destroyed. They thus do not lend themselves as easily to fossilization as do the stronger bones of animals or the hard shells of mollusks.

Birds developed about 180 million years ago during the Jurassic period when the reptiles were reaching their peak. At first they were an inconspicuous new class of vertebrates. Another seemingly unimportant new class of vertebrates developed about the same time—that of small mammals.

Something important happened about 150 million years ago. One day a curious birdlike creature with an unnaturally long tail glided out over the shallow waters of a tropical sea or lake in what today is Bavaria. This creature was about the size of a crow—its plumage may even have been black, but we do not know this for sure. It was possibly flushed out of a palmlike tree that grew on the shores of that warm, tropic sea. Around the shores dinosaurs browsed on shrubs, and overhead large flying reptiles flapped powerful leathery wings—some soared like buzzards in the sky. Our bird was a poor flier, capable only of gliding flight. Well out over the waters and unable to get back to land, it fell into the shallow waters that covered the limy mud flats. There it struggled until it was exhausted and drowned. Time passed, and silt covered the carcass until it disappeared.

Or it may not have been like that at all. This birdlike creature may have died and fallen into a stream and been carried to the sea. Anyway, there it stayed until 1861, when men were quarrying at Solnhofen in Bavaria. The Jurassic rocks of Solnhofen have pure limestone so fine-grained that it has been widely used for lithographic printing. The quarry workers uncovered the excellently preserved skeleton with only the head missing. Here was the find of a century or even of a millennium. "The most precious, the most beautiful and the most interesting of all finds" is how one famous scientist has described *Archaeopteryx lithographica*.

No fossil could ever have been discovered at a more propitious time. With dramatic timing the fossil bird was found only a few years after Charles Darwin's *Origin of Species* had been published and when controversy over evolution was at its peak. Here was a creature that was neither quite bird nor quite reptile—a link in the chain of evolution. Three of the most eminent scientists of the day examined *Archaeopteryx* ("the ancient winged one"). They were the great anatomist, Sir Richard Owen, the first director of the British Museum; T. H. Huxley, Darwin's defender; and Othniel C. Marsh, the famous paleontologist of Yale University. (It's an odd sidelight that this famous fossil was bought by the British Museum for £700 after some haggling.) In 1877 another skeleton, this time with an intact head, was found at Solnhofen. It is now in the Berlin Museum. A third skeleton was found in 1956.*

The eminent scientists pronounced *Archaeopteryx* a bird, although a very primitive one. Indeed, but for the impressions of its plumage, it might almost have been considered an archosaur. (Archosaur, you will remember, is the name of a group that includes the dinosaurs, crocodiles and pterodactyls.) For instance, the tail—only a vestige in modern birds—was very long. The skeleton also had many reptilian characteristics. The jaws were toothed; modern birds have no teeth. But the brain was much larger than in reptiles of this size; *Archaeopteryx* needed a larger brain to cope with the intricacies of flying. But it had a wishbone (found only in birds), and, importantly, the big toe was opposed to the other three toes—a typical bird characteristic and vastly different from a reptile's, which would have had all toes pointing the one way. From an examination of the skeleton it was clear that *Archaeopteryx,* the earliest of all known birds, was capable only of gliding flight. It had no keel on the breastbone on which the great pectoral or flying muscles of modern birds are anchored. Moreover, the wings were short and rounded at the tips. (See Figure 8.)

The most exciting thing about *Archaeopteryx* was its feathers. As paleontologists point out, feathers in all birds are merely modified scales. (And not only are feathers specialized scales, but even the molting each year corresponds to the reptilian sloughing of the skin.) Birds presumably acquired feathers not for purposes of flight—that came later—but for purposes of warmth. Because they trap air, feathers are among one of the finest of all insulating materials. Thus, the acquisition of feathers (possibly from scales that frayed) was of

---

* A poorly preserved fourth specimen was rediscovered in 1970 in the Teyler Museum, Haarlem, the Netherlands, where it had lain unidentified since 1857.

FIG. 8. *Reconstruction of* Archaeopteryx *from Bavarian fossils. Lack of a breastbone indicates it probably glided more than it flapped. Claws on wings aided in climbing. Ancient vegetation is in accord with data at the Smithsonian Institution.*

great assistance to a creature that was acquiring warm-blooded characteristics. By becoming homoiothermic (warm-blooded), with a constant temperature independent of the environment, birds acquired a distinct advantage over their reptilian ancestors. Homoiothermy, too, helped with flight. Flying is strenuous, and homoiothermy helped because it speeded up the body chemistry.

Archaeopteryx's tail feathers lay in rows—they did not radiate from fused tail vertebrae as we find in modern birds. And because the creature had three claws on the end of each wing, it was probably a tree dweller whose wings helped it to glide from tree to tree without coming down to the ground where there was danger from dinosaurs and other reptiles. The young of one living bird, the hoatzin of South America (*Opisthocomus hoazin*) has a wing structure that recalls that of *Archaeopteryx*. The first and second digits have large claws, which help the bird in grasping branches in the fashion of climbing reptiles. Mature hoatzins lose these claws.

Nobody knows exactly how birds acquired flight, but there are two main theories. The first is that the ancestors of the first birds were two-legged, land-dwelling reptiles who raced along the ground and who, over millions of years, acquired wings to take off. The second and more widely accepted theory is that the ancestors of birds were reptiles who lived in trees, and that flight developed because they jumped from branch to branch and subsequently developed parachutelike wings to break their fall. In the course of evolution they developed the ability to glide from branch to branch just as some modern mammals do, including flying squirrels and flying possums (of Australia).

Archaeopteryx is a few rungs up the evolutionary ladder of birds. Scientists think that the ancestors of the first birds were small archosaurs called Pseudosuchia. The Pseudosuchia were thecodonts (the word means "teeth in sockets"), a suborder of small, two-legged reptiles that lived in the Triassic period.

In nature the evolutionary race isn't necessarily won by those who are momentarily superior. For instance, the descendants of the clumsy and comical *Archaeopteryx* ultimately triumphed, while those of the more highly accomplished flying reptiles perished. There was, for instance, the largest of all, *Pteranodon*, with a wingspan of 25 feet, whose way of life may have been that of today's albatrosses. Its fossilized bones have been found in the Niobrara Sea, that great inland sea of North America during the late Cretaceous period.

We don't know why *Pteranodon* and the other pterodactyls became extinct toward the end of the Cretaceous. In very general terms, the birds made better adaptation more swiftly to changing conditions. An exciting pterodactyl fossil recently found in the Soviet Union shows clear evidence of a hairy covering. This suggests strongly that some late pterodactyls were warm-blooded. Moreover, *Pteranodon* was quite well-equipped for flying. In the course of evolution *Pteranodon* had shed as much weight as possible and had hollow wing bones—something that *Archaeopteryx* lacked. In other words, it had become very birdlike.

We can only make informed guesses about the ancestry of birds. No fossils have been discovered showing the progression to *Archaeopteryx*, which shines like a beacon along what someone has called "a very dark road indeed." And the road ahead after *Archaeopteryx* is equally obscure. There is no definite bird skeleton or fragment until the Cretaceous period—a gap of 30 million years. Someday the missing rungs in the evolutionary ladder may be discovered.

By the Cretaceous period, from 135 to 65 million years ago, birds were well established on the earth, with over a dozen genera and two dozen species. Their fossils have been found in North and South America and in Europe. Most of these fossils are of seabirds, which is what you might expect; birds that fall into water have a better chance of becoming fossils (and even then the odds are immeasurably against it) than birds which fall on land where they are devoured by predators. By the Cretaceous period most birds had acquired most of the characteristics of modern birds. They developed hollow bones with air sacs, and also changes in bodily structure. The long, bony tail of *Archaeopteryx* was now suppressed. Some Cretaceous birds have a keel for the anchoring of the powerful pectoral muscle so essential for full flight. But they retained some primitive features; for instance, some still had toothed jaws.

One of the most famous of the Cretaceous birds is *Hesperornis* or "western bird," from a number of fossils discovered in the yellowish deposits of Niobrara chalk in Kansas, U.S.A. This was a large, flightless bird, over six feet long from bill to tail. Its feet and legs were so modified for swimming it is doubtful if it could have walked on land. Its legs protruded so far sideways and were set so far back that on land it probably could have done little more than slither about more or less helplessly on its belly, thrusting at the ground with its legs. In appearance it rather resembled the loons and grebes of today, and it probably specialized in swimming and diving. Its ancestors had flown, but *Hesperornis* had only vestiges of wings. It also had a long, agile neck and long, slim jaws. *Hesperornis* was once thought to have been a toothed bird, and many illustrations show it equipped with sharp, backward-curving teeth. But this is doubtful. One specimen does seem to have teeth, but in others the fossil jawbones are so mixed with other fossilized bones it is difficult to decide. Of possible significance is that a closely related Cretaceous bird, *Hargeria*, which has also been found in chalk pits in Kansas, has no teeth at all. Another well-known Cretaceous bird is *Ichthyornis* or "fish bird," which was about eight inches long and the size of a modern tern. It was presumably a good flier and dived for fish. For a long time it was believed its jaws were toothed, but recent work has shown that what were thought to be the jawbones of one fossil were in fact the bones of a swimming reptile. There is no positive evidence that *Ichthyornis* ever had teeth. Also in Cretaceous times lived a cormorantlike bird; an early flamingolike bird has been found in Scandinavia.

The Cretaceous was the last period of the Mesozoic, the Age of Reptiles. The next era, the Cenozoic, from 69 million years ago to the present, is the Age of Mammals. By the end of the Cretaceous most of the reptiles had perished, including the flying reptiles who had once been masters of the air. The birds, however, continued to develop, and by the early part of the Cenozoic era—the Paleocene and Eocene epochs, from 69 to 35 million years ago—birds had evolved greatly. For instance, many of the modern orders of birds developed. Today we have 27 orders of birds. (Some ornithologists assign birds to 29 orders, putting flamingoes and touracos into orders of their own.) Largest of the living orders is the Passeriformes, which includes flycatchers, larks, swallows, wrens, thrushes, warblers, sparrows, and other perching birds. This large and dominant order has 55 families out of a total of 155, and over 5000 species out of a total of about 8600.

The Paleocene and Eocene epochs saw the arrival of the ancestors of many of the modern orders of birds, including early members of the loons, an ancestral ostrich, the pelicans, ducks, herons, birds of prey, fowls, shorebirds, owls, cranes and others. And by the Oligocene and Miocene epochs, 36 to 26 million years ago, the modern genera of birds had come on the scene.

If you could again take a time machine, this time back to the Miocene epoch, you would not feel you had wandered into too strange a world. For instance, many of the trees and shrubs would look familiar, and so would many of the birds, even if we couldn't identify them precisely. And there would be some that would startle you—birds such as the giant flesh-eaters.

Another exciting discovery was made a few years ago in California, that of a giant oceanic bird from the Miocene epoch. What excited the paleontologists was not the bird's great wingspan of 14 to 16 feet but the toothlike bony projections in both jaws. These "bone teeth" had no evolutionary link with the teeth of *Hesperornis*. Paleontologists think the teeth and jaws had a horny covering.

What was remarkable during the Cenozoic era (which takes in all the epochs we've mentioned) was the development of many huge flightless birds. For instance, *Diatryma*, which roamed the North American plains about 60 million years ago, stood seven feet tall and had a skull as large as that of a modern horse. (Horses of *that* time were about the size of fox terriers.) An almost complete skeleton of *Diatryma* was found some years ago in Wyoming, U.S.A. Its enormous sharp beak and well-muscled legs suggest it was a flesh-eater that ran down small reptiles and mammals. A descendant of *Diatryma* was *Phororhacos* of South America, also flightless and about six feet tall. It looks very much like a monstrous secretary bird! *Diatryma* and *Phororhacos* were possibly ancestors of our cranes and cariamas (South American birds allied to cranes and rails).

There may have been a time when these giant carnivorous flightless birds competed with the early mammals. If this was so, the mammals won the battle for living space, developing into the many modern forms we know now, while these giant birds became extinct.

Both *Diatryma* and *Phororhacos* were presumably descendants of birds that flew. Why did they give up flying? In order to answer that we have to ask ourselves another question and try to answer it: Why did birds develop flight in the first place?

They probably developed flight for reasons of safety as much as for reasons of food gathering. Given a habitat with plenty of food and an absence of flesh-eating mammals, birds will become ground foragers and over the ages lose the power of flight. This happened in New Zealand, the home of the now extinct moas, the largest of which stood ten feet high, and in Madagascar, the home of the extinct ten-foot-high nonflying *Aepyornis*. Hence, plenty of food and an absence of flesh-eating mammals that could prey on them may have been the reason for the development of these two giant flesh-eating birds, *Diatryma* and *Phororhacos,* and also for other large flightless birds that developed later and persist today, such as the rheas of South America, the ostriches of Africa and the emus and cassowaries of Australia. And, of course, there are the penguins. The earliest known penguin fossils date from the Eocene, and some early ones were as tall as a man. Like other flightless birds, their ancestors almost certainly flew. Penguins today are confined to the Southern Hemisphere. One theory about their origin is that penguins were able to abandon flight because mammals did not reach their isolated Antarctic lands, which enjoyed a gentle, temperate climate. The flightless penguins were land dwellers. With the cooling of the polar regions, culminating in the Ice Age, the Antarctic lands no longer could support them. Only one way of escape lay open to them—to take to the seas.

Dr. A. S. Romer makes an interesting speculation in his book *Man and the Vertebrates.* He points out that at the end of the Mesozoic era the giant reptiles had died off and the surface of the earth was open for conquest. The possible successors were two groups—the mammals, from whom we are descended, and the birds. "The former group succeeded but the presence of such forms as *Diatryma* shows that birds were at the beginning their rivals," says Dr. Romer. "What would the earth be like today had the birds won and the mammals vanished?" We wouldn't be here to speculate, of course, and the streams and seas would be in no danger of pollution.

Dr. Romer's is a fascinating speculation, but the great flightless birds had probably already taken a wrong turning. They had acquired useless wings that robbed them of useful forelimbs for grasping and holding. And because lightness was paramount for successful flight, the ancestors of these great flightless birds had not been able to develop large skulls and brains. The future certainly lay with the mammals, which could develop both forelimbs and brains (and, consequently, increasing intelligence).

Here we might take a quick backward look to see why both birds and mammals became warm-blooded.

Mammals developed at about the same time as the dinosaurs, at the end of the Triassic. They were small, inconspicuous creatures, the largest about the size of a cat and most about the size of shrews and mice, and not very efficient. But they had the future before them. Being warm-blooded, they could develop a high metabolism and, consequently, the speed to escape enemies, procure food and reach water. Just as birds developed feathers to conserve heat, mammals developed hairs. Thus birds and mammals could withstand heat and cold much better than reptiles. (Reptiles cannot endure too much heat, and too much cold immobilizes them. Ideally, they need perpetual temperate summer.) And mammals, importantly, incubated their fertilized eggs inside their bodies, unlike the dinosaurs, whose eggs could be predated. Mammals also suckled their young, thus giving them some protection.

This brings us again to the great unsolved puzzle—what happened to the dinosaurs and their fellows, the pterosaurs (flying lizards), the ichthyosaurs (fish lizards) and the aquatic plesiosaurs (sometimes popularly called, not too accurately, swan lizards), which became extinct quite suddenly? In the geological sense of the word "suddenly"—which must be thought of as tens of millions of years—they vanished almost overnight all over the world. "The last of the dinosaurs lived at the very end of Cretaceous times but not beyond that end," says Edwin H. Colbert in *Evolution of the Vertebrates.*

In general terms, the dinosaurs couldn't adapt fast enough to changing conditions, although what these changes were remains obscure. And, as A. Lee McAlister points out in *The History of Life*, separated from the dinosaurs by 70 million years, we may never solve the puzzle. (There's a gleam of hope that new discoveries about continental drift may shed a little more light.)

Again in general terms, the birds and more especially, the humble, protean mammals, because of their advantages of internal temperature control, high metabolism and potential for increasing the size of their brains (and intelligence), could adapt to these obscure but powerful changes.

From the end of the Miocene epoch (about 12 million

years ago) to about a million years ago, there occurred the greatest development and variety of birds. Professor Pierce Brodkorb thinks that about 11,600 species of birds were living together at one time, as against 8600 today (*The Birds*, Life Nature Library). This great bird-population explosion was helped by subtropical climates throughout Europe and North America in the late Miocene and by the spread of seed-bearing plants and particularly grasses and sedges. This rich source of food greatly assisted the evolution of the Passeriformes.

The earth grew cooler during the epoch that followed, the Pliocene, which lasted for ten million years and culminated in the southward march of the glaciers during the Pleistocene or Ice Age. There were, probably, four Ice Ages or periods of glaciation; the last retreated only about 10,000 years ago. These were not as comfortable times for birds as they had experienced earlier, and about one-third of the total species became extinct.

The Pleistocene has provided us with our richest fossil records of birds, notably in asphalt pits in California, in desert lake beds in Kansas and Oregon and in rivers in Florida. (The "asphalt traps" of California were particularly good in preserving not only now extinct birds but extinct mammals, such as wolves, horses, antelopes and saber-toothed tigers. The bones of these creatures, still covered by flesh, were plunged into a liquid that swiftly penetrated their most minute structures and sealed them against most agents of destruction.) One notable Pleistocene bird was a giant vulture aptly named *Teratornis* or "monstrous bird," found in California and Florida. It probably had a wingspan of 12 feet. Even larger was *Teratornis incredibilis*; on the evidence of fragments found in a cave in Nevada, it had a wingspan of about 16½ feet and was apparently the largest flying land bird of all time.

This quick look at birds and their ancestors has dealt with the more physical things—how they may have developed feathers and taken to the air. Along with these physical changes came the development of the nesting instinct and parental care, often by both mother and father. We cannot know how birds acquired the nesting habit. Some experts think it may have developed from movements during mating, and that the collection of nesting material possibly had its origin in birds directing aggressive feelings to grass stems. (We, too, sometimes redirect aggressive feelings to objects; for instance, an angry man may thump a table.) But, of course, these are speculations only. In the chapters to come, we'll look at some of the marvelous nests birds make, the courtship rituals that precede their building, and the migration flights which precede both these activities.

29

MIGRATION

BEFORE BIRDS CAN NEST, male and female must find each other and mate. For most birds the reproductive cycle starts with the springtime migrations.

No spectacle in nature is more mysterious and awe-inspiring than the flight of long skeins of ducks heading north in the Northern Hemisphere. Or you may have heard in the dark night sky the honking of Canada geese heading north to breed. In the Southern Hemisphere I once stood on a rocky headland on the New South Wales coast in late September (spring) and watched a vast gale of soot-colored birds with three-foot-long sickle-shaped wings wheeling down to their age-old breeding grounds in Bass Strait, which separates the Australian mainland from Tasmania. These were short-tailed shearwaters (*Puffinus tenuirostris*, from Latin *tenuis* = slender, and *rostrum* = bill), members of the petrel family. They were strung out in a great palpitating cloud stretching as far as I could see over the Pacific. For half an hour the birds, known popularly as muttonbirds because of their tasty flesh, sliced their way past

249

me, heading for their rookeries on the coasts of Victoria, South Australia and Tasmania, Phillip Island and the islands of the Furneaux group (with such picturesque names as Big Dog, Little Dog and Babel).

Such a spectacle fills you with wonder at the prodigality of nature. It is even more overwhelming if you should chance to be on a breeding islet; then your mind is stunned by the immensity of numbers that spread a huge, quivering umbrella over you—as was the early navigator Matthew Flinders in 1790. One dawn Flinders saw "a stream of sooty petrels, 50 to 80 yards in depth and 300 yards or more in breadth, passing without a stop for a full hour and a half." * He estimated the number at 151.5 million. Incidentally, Flinders came upon some small islets inhabited by geese, shags, penguins and sooty petrels, "each occupying its separate district, and using its own language. It was the confusion of noises among these various birds which induced me to give the name Babel Islets to this small cluster."

The birds that Matthew Flinders—and I—saw were on the last stage of a nine-month journey, in which they had made a great figure eight of 20,000 miles in the Pacific, going as far north as the Bering Sea before heading south down the coast of California.

Why do they go so far? Probably it is the need for food that drives them—for krill (small shrimp), anchovies and cephalopods (mollusks).

Food, too, is probably one of the basic causes of all other great bird migrations. The others are climate (warmth and humidity), light and habitat. And all, as you've probably observed, are interrelated.

A bird that flies north in the northern summer to breed continues to live in the temperature that suits it best. A striking instance is the ruby-throated hummingbird (*Archilo-*

---

* *A Voyage to Terra Australis*, 2 vols., 1814.

chus colubris), which winters in Central America and advances at about the same speed as the 35° Centigrade (89.6° F.) isotherm. (An isotherm is an imaginary line joining areas with an equal mean temperature. An isotherm moves north in the Northern Hemisphere with the advance of summer.) The increase in temperature stimulates the opening of the flowers on which the ruby-throated hummingbird feeds.

The ceaseless search for food for themselves and their offspring dominates the lives of all birds. Birds have conquered the air and gained a mobility no other creatures possess. But they have had to pay a heavy price. Flight is achieved only by a high metabolism. The body temperatures of birds are six to twelve degrees higher than those of mammals. Constant refueling is therefore needed to maintain the bird's body temperature and its constant activity and high energy output. Every day a small bird must find at least one-third of its weight in food. Translate that into human terms and you'd have a 168-pound man compelled to eat 56 pounds of food a day! A chickadee has been known to eat over 5500 canker-worms a day, and a woodpecker to swallow 3000 ants in a single day. An owl, in order to slumber peacefully during the day, was observed to hunt all night to capture ten mice.

Thus birds work hard for their living, with death from starvation always facing them. Ornithologists think that more birds die from starvation than from any other single cause.

The bird's day begins at sunrise and ends at nightfall. During the day it may take short rests or even "catnaps." These are essential because it's a lucky bird who isn't disturbed at night by a predator—even if the bird isn't the immediate quarry.

Fortunately for birds, their metabolism drops sharply at night, and with it the internal temperature. A bird with a day temperature of 108° will have one of 68° or thereabouts at night, with a considerable slowing down of its bodily chemical processes. Thus assisted, a bird can survive so many consecutive hours without eating. This is a natural but not a simple type of sleep; it reaches its extreme form in those animals that hibernate.

Without this adaptation it's doubtful if birds—and particularly hummingbirds—could survive the night. No bird has a higher metabolism than a hummingbird. Its energy output when it's feeding is about ten times that of a man running at nine miles per hour. To give out the same amount of energy, a man would have to eat 150,000 calories a day—about 50 times his normal consumption.

When migrating, most birds break their journey into spells of flying and resting. Small birds generally travel about 50 miles a day. Larger birds can cover greater distances. A blue-winged teal (*Anas discors*) was banded in Minnesota and recovered ten days later 3000 miles away in Colombia, South America. Some of the larger birds perform remarkable feats of nonstop long-distance flying. The Atlantic golden plover (*Pluvialis dominica*) makes a nonstop or almost nonstop flight of 2500 miles from Newfoundland to Colombia. Then it pushes on, more leisurely, to Argentina. The Pacific golden plover flies almost 3000 miles across water from Alaska to Hawaii. The black-poll warbler (*Dendroica striata*), which passes the summer in the Yukon, makes a 4000-mile trip with stops back to Brazil. Some of the smaller land birds feed by day and fly by night; others fly by day, but too high to be seen from the ground. Most of the larger birds fly by day. Speeds range from 20 miles per hour for smaller birds to 60 miles per hour for ducks and swifts. The Arctic tern (*Sterna paradisaea*) is undoubtedly the greatest long-distance flier. This sea swallow avails itself of the best of both worlds, spending one summer in the Arctic (where it breeds) and the following one in the Antarctic. Pursuing the polar summers, it commutes about 22,000 miles a year. This strenuous life of the Arctic tern is one of the greatest puzzles of migration—which, in truth, is already mysterious enough. Why should the bird fly so far? One possible explanation is that during the Ice Ages the birds migrated much shorter distances, and then, as the ice receded, had to increase their distances considerably.

Some birds breed in Greenland, Iceland and Lapland, but spend the winter in the temperate regions of the United States.

Besides the north and south migrations, there are vertical migrations. For instance, in North America the gray-headed junco (*Junco caniceps*), the pine grosbeak (*Pinicola enucleator*) and the black-capped rosy finch (*Leucosticte australis*) spend their summers on mountaintops and their winters in the valleys and lowlands. An interesting east–west migration is that of a small shrike that nests in Central Asia and winters in Africa—2500 miles west and only 1200 miles south. Often birds of the same species, such as the song sparrow (*Melospiza melodia*), may be permanent residents in one area and migrants (or nomads) in another. Others again may be irregular migrants, only leaving a region when food grows scarce.

Although migration brings gains for birds in climate, food and nesting sites, it is also hazardous. It takes a fantastic toll of birds. Chief among the causes of death are lighthouses, snow, sudden coldsnaps, gales, drought, floods, grass and forest fires and man. A late blizzard in the north central United States a few years ago killed over a million small birds. Tens of thousands of short-tailed shearwaters are sometimes washed ashore dead on Australia's southern coasts, unable to make their landing to breed in Bass Strait.

Indeed, so dangerous is migration that Dr. David Lack suggests that the reason why the European robin (*Erithacus rubecula*) winters in England and doesn't migrate is because the hazards of the English winter are no greater than the bird would encounter in migrating.

How birds navigate does not greatly concern us in this book. It remains one of the great mysteries, although recent research has opened the door a little; we know that some birds presumably steer by the sun or stars. For instance, a Manx shearwater (*Puffinus puffinus*), taken to Boston from its nest burrow in Skokholm, Wales, and then released, was back home, 3000 miles away, in twelve and a half days.

Experiments with Adélie penguins (*Pygoscelis adeliae*) show that they take their bearings from the sun and find their way back to their rookery when released up to 2000 miles away. When released they swim northeast, but when the sun is obscured they swim at random. Wherever the penguins are on the Antarctic continent, the sea always lies to the north. They presumably take a northeasterly bearing to counteract the westerly current. Two penguins set down on Wilkes were back at their home rookery at McMurdo, 2000 miles away, in ten months. They had averaged a remarkable speed of eight miles per hour.

Our major concern here has not been with the "how" but with the "why" of migration, and the answer, as we have seen, is based on food supply for offspring and increased hours of daylight in which to gather it.

In the following chapter we'll look at a typical or standard bird that migrates to a suitable place to raise a family. As we'll see, it is usually the male who initiates the raising of a family.

30

LANDOWNERS AND COURTSHIP

LET US IMAGINE that we have a movie camera that can focus, first of all, on a mixed flock of typical or standard birds in winter, and then on a single bird in that flock. Our typical single bird is a male. He is also a land bird and a passerine or perching bird and frequents hedges, gardens, scrubland and woods. He is a songbird—although he sings little or not at all in winter—and he is usually monogamous. He eats insects and seeds. He is a resident in these autumn and winter quarters, living in a mixed flock of males and females.

There are special reasons for choosing a perching bird and a songbird. The order Passeriformes, that of perching birds, is the largest of the class Aves; it has about 5100 species, over half of the known 8600 species of birds. They all have feet adapted for perching—three toes point forward and one backwards and they lock in a grip that cannot be released until the bird takes wing. When the bird's feet grip the perch, the bird's weight tautens a tendon that locks the foot around the perch. This enables a bird to sleep on its feet.

Songbirds make up a suborder, Oscines, of the Passeriformes. This suborder is based not so much on the bird's
singing ability but on the structure of the syrinx, a complex
apparatus of muscle fibers in which the voice is produced.
Some songbirds, in fact, do not have musical voices; these
include the harsh-voiced crows, raucous jays, chattering
magpies and birds of paradise.

Our typical or standard bird is thus a songbird, in the
autumn and early winter an amicable member of a mixed
flock with whom he hunts cooperatively for food. But if we
return with our camera in the late winter, we'll note that he
has become quarrelsome—and so have his fellows. He pecks at
other males, and they at him. Or he may raise his wings and
assume an aggressive posture and chase another bird. We'll
notice, too, that he is beginning to grow his courting plumage
—he was as dull as the females in midwinter.

The flock, once so friendly and cooperative, is beginning
to break up. Our male bird sometimes sings snatches of song.
What is happening to his body has been influencing his
behavior. His testes or gonads (sex organs), which had shrunk
to minuteness during the autumn and winter, are increasing
rapidly in size. They may expand as much as several hundred
times in size.

One day our male leaves the flock and flies off, perhaps
hundreds of miles north, driven by instinct to find a piece of
land or territory he can claim as his own, and where he can
advertise for a wife. Because he is a songbird, he'll do it with
song.

After his migratory flight, our typical bird finds a suitable
place—sometimes he returns to the very place where he mated
last year. There, from a vantage point such as a tree or even a
housetop, he proclaims in a very loud voice. He is warning
other males that this is his own piece of real estate, on which
he proposes to raise a family, and is also telling any females of

his species in the neighborhood that a well-endowed bachelor is in the marriage market. Occasionally another male of his species may seek to invade his piece of land, and then ensues what British ornithologist James Fisher in *Watching Birds* has aptly called a game of "song-tennis," with both birds hurling full-throated and melodious challenges at each other over the neutral ground between them. In time the games of song-tennis played by our typical bird and other males around him are resolved, each of them agreeing to allow the other to establish himself as the owner of a particular plot of land. The boundaries have been defined and come to be respected. Thus birdsong is of great biological value in avoiding conflicts.

Singing from a full throat, our typical bird will finally lure a bride to the area of which he has taken possession.

One of the most important functions of territory is that it spaces birds out evenly during the breeding season so that all parents and their offspring are assured of a supply of food. It could be argued that songbirds are a step ahead of man in resolving most of their territorial differences and sexual competitiveness mainly by musical contests. The differences are resolved usually bloodlessly because it has been found that the further an intruder flies into another's territory, the less aggressive the intruder becomes and the more aggressive the owner becomes. Only on the borders are both birds equally matched.

When our standard bird started to engage in song-tennis he had probably staked out about two acres. But when all the song games have been resolved, he may own an acre or even half an acre. Territories among songbirds may range from about half an acre to as much as four and a half acres. For instance, an American robin (*Turdus migratorius*) may want half an acre or less, while a western meadowlark (*Sturnella neglecta*) may require up to 20 acres. The average is generally about an acre. Among birds other than songbirds there is even wider

variation. A black-headed gull (*Larus ridibundus*) may be
content with about three square feet—the immediate vicinity
of its nest—but a golden eagle (*Aquila chrysaetos*) may want six
square miles.

Although our typical bird will defend his piece of land
with song and by chasing other males of his species, he will not
generally attack birds of other species, even those that eat the
same food. I suppose in this he's not so very different from
ourselves. If a stray cat wanders across our lawn, we are
generally tolerant. It is only when another person trespasses
that we are liable to take offense.

Territory may thus be defined very simply as any area
that a bird defends against its own kind. Other more
specialized definitions are used by ornithologists, but these
need not concern us here.

Like much that birds do, this song-tennis is innate
behavior; the male bird presumably has no idea that he is
warning off other males and advertising for a bride. He does it
because he has to do it. For centuries man has listened to the
melodious songs of birds and, projecting his own emotions into
birds, has thought that birds sang because they were in love
and because they were happy. This is, of course, expressed in
popular sayings, such as "happy as a bird" or "merry as a
lark." Another popular error has been made by poets and
storytellers, who usually attribute song to the hen bird. The
poets have been strong on imagination but weak on ornithology. The female nightingale (*Luscinia megarhynchos*) does in fact
sing. So do some other female songbirds, but they are not as
good or consistent singers as males.

The reasons for birdsong are bread-and-butter ones,
although the products are probably pleasing to female birds as
well as to ourselves. Musical appreciation is probably innate in
all animal creation. (Charles Darwin thought that our early
ancestors probably used musical tones as a means of courtship

and, moreover, that they probably did so before they learned to speak.)

In passing, the best singers are usually small dull-colored birds that live in thick scrub or forests, where sound is of paramount value in maintaining territory against other males and luring brides. (In open country, vision may suffice and allow birds to avoid each other—many open-country birds are conspicuously colored.) Some of the more melodious feathered minstrels are so leather-lunged that you can hear them half a mile away. (This has proved an additional form of protection for some birds, because no wants a captive bird that can deafen him.) Another reason why songbirds should be dull-colored and blend with their surroundings is that birds who sing are vulnerable; they advertise their presence to enemies. The nightingale, one of the most famous of the songbirds, is typically a small, rather dull bird. Although it is associated in our minds with England, it could be equally linked with Africa because it winters on that continent. It is, of course, a splendid singer, but some of its fame is certainly due to having had the best public-relations staff—the great English poets.

Three noted North American minstrels are also rather drab birds. These are members of the thrush family—the wood thrush (*Hylocichla mustelina*), the veery (*Catharus fuscescens*) and the hermit thrush (*C. guttata*). It was the wood thrush that stirred Thoreau to this tribute:

> The wood-thrush's is no opera music. It is not so much the composition as the strain, the tone, that interests us—cool bars of melody from the atmosphere of everlasting morning and evening. . . . The thrush alone declares the immortal wealth and vigor that is in the forest. Here is a bird in whose strain the story is told. Whenever a man hears it, he is young and nature is in her spring; whenever he hears it, there is a new world and one country, and the gates of heaven are not shut against him.

Thus we have our typical or standard bird—a songbird.

Although the purpose of birdsong has been fully understood only in the last 50 years or so, the concept does in fact have quite an ancient history. For instance, Aristotle (about 350 B.C.) wrote, "A pair of eagles demands a large space for its maintenance and on that account cannot allow other eagles to quarter themselves in the close neighborhood." A hundred years later Zenodotus noticed that "one bush does not shelter two robins." After the great Greeks there were no acute observations until 1622 when G. P. Olina of Italy observed the European nightingale's "first coming to occupy or seize upon one place as its Freehold, into which it will not admit any other Nightingale but its mates." There were other isolated observations before the famous English naturalist Gilbert White of Selborne wrote in a letter in 1772: "During the amorous season such a jealousy prevails amongst male birds that they can hardly bear to be together in the same hedge or field. . . . It is to this spirit of jealousy that I chiefly attribute the equal dispersion of birds in the spring over the face of the country."

The first comprehensive account of territory came in 1868 when a German ornithologist, Bernard Altum, published his book *Der Vogel und Sein Leben* (*The Life of the Bird*). Born in 1824 in Munster, Westphalia, Altum became a Roman Catholic priest but took such an interest in birds that at the age of 31 he returned to his university to study sciences. He rapidly became one of Germany's leading ornithologists. Altum pointed out that it was impossible among a great many species of birds for numerous pairs to nest closely together. Individual pairs, he said, must settle at precisely fixed distances from each other. The reason for this necessity was the amount and kind of food they had to gather together with the methods by which they secured it. Birds, he said, needed a territory of definite size, which varied according to the productivity of a given locality. He pointed out that song was used both as proclamation of territorial boundaries and also as an invitation to females. This account of territory was widely accepted in Germany after an initial controversy, but won little acceptance abroad, probably because of translation difficulties. However, Altum was not the man to win the admiration of sentimental bird lovers who gushed over the devotion of birds for fledglings. He writes: "The so-called love of offspring is the urge to feed birds of such and such a shape, uttering such and such cries, fluttering their wings in such and such a manner and opening their beaks wide, but it is not love." This explanation would be accepted by most ornithologists today.

The mainspring of most of the modern study of territory —not only in birds but in other creatures—stems from the publication in 1907 of a book called *The British Warblers,* by an English amateur naturalist, H. Eliot Howard. This modern exposition on territory was expanded on in another book by Howard in 1920—*Territory in Bird Life.* This is one of the germinal books in the study of ornithology—and a delightfully written one. (I have a copy on my bookshelves and frequently dip into it, not so much for Howard's observations but in order to share his delight in studying birds and to savor the beauty of his style.) Born in 1873, Howard was a steelmaster, a director of a great English firm, and a dedicated naturalist. This "quiet genius" spent hours in the field observing and recording the behavior of his beloved birds before departing to his office.

Some birds defend their territory not with singing but with threat displays. These may include aggressive flights, which are more bluff than anything else, or, like yellow wagtails (*Motacilla flava*), they may threaten each other by puffing out their bright-yellow breasts at each other, rather like small pouter pigeons. Male European robins display their red breasts to each other. Red has much the same effect on robins as it is alleged to have with bulls. Dr. David Lack found that a tuft of red feathers placed in the territory of a male robin was enough to stir the bird into vigorous posturing or even to an attack, while a stuffed robin, colored brown like an immature robin, was ignored.

When songbirds fight with sound, the winner is usually the one with the best lungs. The loser retreats. (Once as an experiment I played at full volume a recording of the male challenge call of a magpie lark [*Grallina cyanoleuca*]. The startled male who thought he owned my garden and my neighbor's shot off at full speed.) However, savage fights to the death can occur among some birds. White storks (*Ciconia ciconia*) fight so savagely for territories that sometimes eggs are destroyed or one of the birds killed.

Penguins, too, are savage in their territorial fights. They rain blows on each other with their flippers, and jab and bite with their powerful bills. Professor Carl Welty thinks that this is probably because of the shortage of suitable nesting sites. Moreover, nesting materials, such as pebbles, are sometimes in short supply, and penguins in colonies may spend a lot of time in stealing from each other. Penguins often have an eye missing as a result of territorial or courtship battles. The fighting for burrows among two or more pairs of Peruvian penguins (*Spheniscus humboldti*) continues until all are blood-smeared.

We have looked at the behavior of a typical perching bird in finding a territory and defending it against other males. But this behavior is not necessarily typical of all birds. As we have seen, the average of half an acre of territory is considerably reduced among some social birds, such as seabirds, which defend the area only immediately around the nesting site, and it is greatly expanded among birds of prey, such as eagles, which may maintain six square miles.

In addition, there is another and quite remarkable form

of territory among birds known as "lek birds." "Leks" are small sites to which these birds go only for courtship, and have no connection with the nest site or with the raising of a family. These are display arenas and courts where the males display, sometimes alone or sometimes to females, and each male stakes out and defends his own. "Lek" is apparently derived from Swedish *leka* meaning "to play," which can have a sexual meaning. The courtship behavior of lek birds is so remarkable and entrancing that it deserves at least a full chapter of its own. We've given it two, following this one.

Meanwhile, where is the bride of our typical bird all this time? She generally leaves the mixed flock some time after the male has left and in her turn starts her migration. In time she arrives at the place, sometimes hundreds of miles from where she started, where males of her species are already claiming their territories. She may fly over a number of pieces of real estate in which males are advertising for brides, because she is not yet ready to mate. But in the ripeness of time, as her sexual organs increase in size and produce small eggs, not as yet coated with shell, she will fly down to join an importuning male. What follows is a fascinating episode whose mysteries have been unlocked by the work of various naturalists, and most notably by Niko Tinbergen and James Fisher. At first the female may be received in unfriendly fashion; the male sometimes thinks she is another male invading his property and he greets her with challenging bursts of song and with pseudoattacks, fluttering his wings, flying close to her, soaring in the air. Suddenly he recognizes her, and what may be described as an aerial courtship or sexual chase ensues. He pursues her round and round his territory, trying to bring her to the ground so that he may mate with her. More often than not she is not as yet ready to mate, and although she will now stay in the territory, days or even weeks may pass before she is ready.

On the next day and probably for days afterward there are more sexual chases around the territory. From time to time the female may wander away from the territory, only to be lured back by the male's singing. As the days pass she wanders less and less, and shortly she joins her spouse to be in the defense of the territory. She is not yet ready to mate but she attacks not other males but females. One day when her breeding organs are ready, she no longer flees from the male. This time, when he goes into his courtship routine, she invites him with flattened back and lifted tail. The mating of birds is always a delicate matter in which both birds must cooperate fully. Birds have no penises (with the exception of some water birds), and union is achieved by the placing of both cloacas (or vents) together. They are now ready to begin nest building.

Our typical bird courted his bride with song. Some other birds woo with a variety of means, such as aerial flights, dancing, and feeding the female.

Some aerial wooings are spectacular. For instance, the great spotted woodpecker (*Dendrocopus major*) hovers before his mate to show her the fiery-red feathers under his tail. The male lapwing (*Vanellus vanellus*) soars high above the ground, propelling himself at a sharp angle. At the apex he plunges suddenly downward, turning, twisting and somersaulting like a stalling aircraft. When he comes out of the dive he veers off erratically, the whole time producing a loud humming throb with his wings. Other birds court with dancing, notably cranes, where male and female may perform a stately gavottelike dance; a pair of males may dance to an audience of females, or they may even do so in mixed flocks.

Dancing is not confined to land birds. Some birds perform water dances, particularly the great crested grebe (*Podiceps cristatus*). Male and female swim toward each other and touch beaks; then they dive and surface to face each other, each holding a piece of waterweed in its bill. Or they may perform the so-called penguin dance, in which both birds, breast to breast, rise above the water, paddling energetically to stay up while their heads sway from side to side. Razor-billed auks (*Alca torda*), called razorbills in Europe and so named because of their sharp bills, have a communal paddle dance on the surface of the sea. Among these birds, dancing appears to have developed beyond mating into communal fun. They move first in single file, and then form a circle, converging until their beaks almost touch. The circle swells and then breaks. Each pair of birds bow and come together and hold beaks and waltz around each other. Then they form a single line with beaks and tails raised. Then, once more, they make a ring before breaking with the pairs facing each other and waltzing around each other. As one observer puts it, "It is as good a description of a folk dance as any I have heard of. And what is sex to start with ends up, it seems, as pure enjoyment of rhythmic patterns of motion."

Other birds, again, may court with colorful ceremonies that have developed from their daily life. The male may feed the female, as is the custom with robins, gulls and parrots. Or, like gannets or smews, they may go through the motions of drinking although they take in no water. Or they may specialize in presenting and exchanging nest material with each other.

Among most birds the male takes the initiative in courtship, but it is mutual among many seabirds, and may even be social, with many birds involved, as we shall see in the chapters on lek birds. Among other birds, such as many of the species of ducks, the female may take the initiative and among common terns (*Sterna hirundo*) either male or female may take the initiative.

Whatever form the courtship takes, with most birds it serves the purpose of strengthening the bond between male and female so that they will carry out their tasks of reproducing themselves—tasks such as nest building and caring for the young.

Most birds are monogamous, and the pair bond may be for life, or several years, or a year, or perhaps just for this particular brood of young. Adélie penguins are thought in many cases to pair for life. Life "marriages" are thought to persist among a number of seabirds, including royal albatrosses (*Diomedea epomophora*), Manx shearwaters and the British storm petrels (*Hydrobates pelagicus*).

The bond between two birds can be very strong. For instance, on two occasions the partners of shot black ducks (*Anas rubripes*) refused to take wing and leave their dying mates when the other ducks flew away from the guns.

Although it is not as common as monogamy, polygamy occurs among many birds. The male may mate with several females, or one female may mate with several males. Polygyny, in which one male mates with a number of females, occurs regularly in many gallinaceous species, such as pheasants. A notable example is the blue peafowl (*Pavo cristatus*). Ostriches and rheas are other examples.

Polyandry, in which one female is mated to several males, is most common among those birds in which the males hatch the eggs and rear the young. These include the tinamous (Tinamidae), jacanas (Jacanidae), painted snipe (Rostratulidae) and button quail (Turnicidae). The males of these families can be described as "henpecked"; the females are bigger, wear the gay courtship feathers, do the courting and defend the territory. The drab males build the nest, incubate the eggs and rear the young.

The formation of a bond between male and female birds is a most important adaptation which makes possible among many birds the cooperative sharing of the task of raising a family. Man apart, there are few animals in which this cooperative behavior is as highly developed as it is among birds. Among many birds the ceremonies of courtship do not end with mating but persist afterward and, most importantly, help to reinforce the pair bond. These postcourtship ceremonies sometimes take marvelous and beautiful forms, such as among the little egrets (*Egretta garzetta*), which Sir Julian Huxley observed in Louisiana and of which he has written so engagingly:

> Some little time before the human watcher notes the other's approach, the waiting bird rises on its branch, arches and spreads its wings, lifts its aigrettes into a fan and its head-plumes into a crown, bristles up the feathers of its neck, and emits again and again a hoarse cry. The other approaches, settles in the branches nearby, puts itself into a similar position, and advances towards its mate; and after a short excited space they settle down close together. This type of greeting is repeated every day until the young leave the nest; for after the eggs are laid both sexes brood, and there is a nest-relief four times in every twenty-four hours. Each time the same attitudes, the same cries, the same excitement; only now at the end of it all, one steps off the nest, the other on. One might suppose that this closed the performance. But no: the bird that has been relieved is still apparently animated by stores of unexpended emotion; it searches about for a twig, breaks it off or picks it up, and returns with it in its beak to present to the other. During the presentation the greeting ceremony is again gone through; after each relief the whole business of presentation and greeting may be repeated two, or four, or up even to ten or eleven times before the free bird flies away.*

As Sir Julian observes, the sight of a pair of egrets changing places on the nest, bodies bowed forward, plumes a golden fan of lace, absolute whiteness of plumage relieved by gold of eye and lore and black of bill is a sight that no one can ever forget. Such unforgettable scenes are not confined to the little egrets;

---

* "An Essay on Bird Mind," in *Great Essays in Science*, Martin Gardner, ed. (New York: Washington Square Press, 1962).

the crested grebe, which we looked at earlier, has elaborate and beautiful, highly charged emotional ceremonies where one bird leaves the nest and greets its mate and both perform what can only be described as aquatic dances, one bird sometimes diving under the other.

Even those we may regard as unlovely birds, such as the king shags (*Phalacrocorax albiventes*), bill and coo; both birds bow, kiss and nibble each other about the head, meanwhile uttering grunts and murmurs of affection.

The bird world is nowhere more marvelous or beautiful than in its courtship and mating ceremonies. We cannot fail to see analogies with our own behavior. For instance, courting couples indulge in baby talk much as some courting birds do. Dr. Konrad Lorenz, in *King Solomon's Ring,* has reported that every delicacy the male jackdaw finds is presented to his bride, and that she accepts it with the plaintive notes typical of baby birds. "The love whispers of the couple consist chiefly of infantile sounds," says Dr. Lorenz. The only difference between a courting human couple and the jackdaws would appear to be that the jackdaws don't reflect on what they are doing and that the human pair, if it did so, doesn't care.

Another amusing similarity is that newly mated geese behave much the same way as newly engaged couples do. Mating is not just decided by pairing off but by a very special triumphant noise that the gander makes after he has driven off the rival—either a real or imaginary one. Only if the goose is willing to accept the proposal does she join in this triumphant proclamation. As Professor Oskar Heinroth has commented, it is as though the birds profess their affection for one another by facing the whole world together! Moreover, when the courting gander approaches his mate to be, it is as though he were trying to ingratiate himself by running down others and glorifying his own heroism! Professor Heinroth says, much to the point: "For me the behaviour of geese is not a sign of human intelligence, but merely shows that many of the things we do, such as the running down of others and the glorification of ourselves, are simply social instincts. An individual does not necessarily think about them. Very often they hardly lend themselves to analysis." *

* Quoted in David Katz, *Animals and Men* (London: Pelican, 1953).

31

MALE DANCERS AND SINGERS: PART ONE

MOST MALE LEK BIRDS who seek to win brides with dancing displays or vocal appeals are usually equipped with fine feathers, or striking ways of producing sound—or both. The lek birds you are likely to be most familiar with are the North American prairie chickens or grouse—the greater prairie chicken (*Tympanuchus cupido*) and the lesser prairie chicken (*T. pallidicinctus*). As winter departs the cocks begin gathering on their arena or booming ground. (The arena of the lesser prairie chicken is usually called a "gobbling ground.") Starting well before sunrise, the males strut, erect the feathers on their neck, inflate their orange neck sacs and then let out their sacs with a resonant boom. They perform for an hour or more before dispersing, to reassemble once more in the evening. At the first gatherings there is some displaying and fighting with other males, but, as the days pass, each cock establishes himself as the owner of his own court. It starts as an all-male show. The females come later, usually in March, and they move around in loose flocks while the males

270

inflate and deflate their large orange sacs. Mating occurs when a captivated female enters the court of a male. After some days the female departs to raise her family alone.

Some of the booming and gobbling grounds are centuries old, and the birds persist in returning to them even when a railroad or highway is built across them.

Even more spectacular but less noisy are the arenas of the sharp-tailed grouse (*Pedioecetes phasianellus*), of the open grasslands of North America. In the breeding season, the males have orange or lavender air sacs. The cocks all dance together, stamping their feet rapidly with their heads down and tails up. When one slows his step, they all slow, and when one stops, they all stop, in perfect timing. The dances of the prairie chickens and the sharp-tailed grouse were the inspiration of some of the dances of the plains Indians—just as half a world away some of the African tribes adapted into their dances the courtship ceremonies of the ostrich (*Struthio camelus*).

The sage grouse (*Centrocercus urophasianus*), which lives on the sagebrush plains of western North America, sometimes have immense arenas, half a mile long and 200 yards wide, where up to 400 cocks compete with each other for females. They have been closely studied in Wyoming by two U.S. ornithologists, John Scott and James Simon. In the mating season the sage-grouse cocks have greenish-yellow air sacs and wattles. After some strenuous weeks of display and fighting, the birds sort themselves out into a hierarchy, with one bird so dominant over the rest by the vigor of his strutting and calling that he becomes the consort of almost all the females. They make their way to the court of this athletic overlord, passing through the courts of lesser birds who make no move to mate with them. The number-two bird in the hierarchy has his court close to the dominant bird, and his strutting attractions are almost as winning as those of the overlord. (See Figure 9.)

In Europe, the black grouse (*Lyrurus tetrix*) also has

FIG. 9.   *Sage grouse on their booming grounds.*

dancing grounds where the males perform stylized jumping dances while uttering loud hooting cries. They assemble each spring on ancient leks, spread themselves out, and each cock begins jumping up and down, hooting all the while, on a selected spot. The male continues to do this until another approaches his bit of real estate. Then the owner thrusts his head down and neck forward, his red eye combs swollen with blood, and races towards the intruder. He then retreats, and the intruder likewise retreats. The fighting, as in most fights between birds, is highly ritualized.

The courtier of the lek world is the ruff (*Philomachus pugnax*), a small sandpiper of Europe and Asia. The breeding plumage of each bird includes a tiny ruff, rather like that of a

Dutch burgher or Elizabethan courtier. No two ruffs are alike.
They may be white, black, buff, purple, red or even ginger and
they may be plain or barred; the combinations are endless.
Ruffs pass the winter in Africa. Then, early each spring, the
males return to their arenas, where 20 or 30 of them assemble
and gradually agree to live and let live. Unlike the prairie
chickens, their displays are silent. At first the males display
only to each other, darting in all directions with the erected
ruff. Suddenly, after a series of these erratic sorties, the male
stops abruptly and, with wings half-open and ruff erected,
sinks slowly down with his head bowed until his beak touches
the ground. The whole time the bird is quivering as though in
the grip of an intense ecstasy.

When the females arrive the displays become even more excited. The birds rush wildly and threaten each other even more intensely before each male in turn sinks slowly down with head bowed until its beak is thrust into the ground. We cannot know what goes on in the minds of the plain little females (called "reeves") while they inspect these curtseying gallants. They don't appear to be unduly excited. It's possible that they choose the cocks with the most beautiful ruffs; when a female does make her choice she nibbles his ruff or neck feathers. Then they go off together to mate. No one interferes, although the courts are only a foot or two from each other.

After mating, the ruff and reeve separate and possibly may never see each other again. Males and females are believed to live in separate flocks during the rest of the year and to meet only for mating.

Some tropical birds, among them the most resplendent dandies of the bird world, hold court in forest arenas. They include the birds of paradise of New Guinea and surrounding islands, the argus pheasants of Asia, the bright-orange cock of the rock of South America, the tiny jeweled manakins of the tropical Americas, and a number of hummingbirds. In the following pages we'll take a look at some of them.

The small magnificent bird of paradise (*Diphyllodes magnificus*), which has a red back, a glossy golden ruff, an iridescent green breast shield and, to cap it all, two long, thin, curled tail feathers, clears an area on the ground for its court. On the other hand, the thrush-size King of Saxony bird of paradise (*Pteridophora alberti*), which is almost certainly the loveliest of the 60-odd species, displays on a branch high in the trees. When R. Bowdler-Sharpe, a famous British ornithologist, first saw a made-on-the-spot drawing of the King of Saxony bird of paradise he exclaimed, "It is impossible that such a bird can exist in nature!" That is a valid reaction to one's first sight of this bird, who appears to wear an enameled lyre on his head.

Poised on each side of his head are long shaftlike plumes with a series of little blue flags that shine as though they are made of enamel. The stiff plumes coil like fern fronds at the end and may be as much as 18 inches long. The opinion of some naturalists that the King of Saxony bird of paradise is the most beautiful of the species known to science today is also echoed by discriminating natives of highland New Guinea, who prize its blue-enameled plumes above all others. In New Guinea, man's status is denoted not by Cadillacs and swimming pools but by the number and beauty of the bird-of-paradise plumes he flaunts.

Only the male birds of paradise have the fine plumage. The females are usually drab creatures; bright feathers would only bring predators to the nest. Male finery is thus for courtship only. That apart, it is a liability; the long, trailing plumes of some species impede flight, and the handsome ruffs, breast plates, capes and headdresses of others betray them to enemies.

Courtship among such gorgeous gallants is often very competitive, with males putting on sustained song-and-dance displays. The magnificent bird of paradise displays with each male bird out of the sight of the others. But the great birds of paradise (*Paradisaea apoda*) compete with each other on neighboring tree boughs. Often, three or four males will perform in turn to the watching ladies, dancing and gyrating, erecting and shaking out their plumes until they look like animated jewels. Some birds of paradise who court on tree branches climax their dance with a daring gymnastic feat of hanging upside down while the light glitters from their iridescent, quivering plumage.

After the birds mate, the female goes off alone to lay the eggs and raise her family alone; the gaudy father, sensibly we might think, stays away.

One of the most acrobatic of the birds of paradise is the

male standard-wing bird of paradise (*Semioptera wallacei*), who performs a backward somersault from his perch and lands on the ground with his wings closed.

Birds of paradise have captured Western man's imagination since 1522 when members of Magellan's expedition, which had circled the globe for the first time, brought the first dead specimens to Europe. (Indian princes knew the plumes long before this and treasured them.) After Magellan's death off the Philippines in April, 1521, the expedition went to the Moluccas where the Rajah of Bachian presented the Spaniards with some dead birds of paradise. These had been prepared in the usual Malayan manner: the entrails had been taken out and the legs and wings cut off to prevent decay. The long plumes of the birds, which are purely ornamental and play no part in flight, grow in long flowing streamers from underneath the wings.

The birds when presented to the King of Spain caused a sensation. The Malays had called them *manuk dewata*, or birds of the gods. Antonio Pigafetti who wrote an account of the Magellan expedition recorded the gift and embarked on a little fancy. He wrote: "These birds have no wings but instead of them, long feathers of different colors like plumes. They never fly except when the wind blows." Pigafetti gave the birds feet—"legs slender like a writing pen and a span in length." He had seen the prepared birds and had made an honest mistake of thinking that the plumes were capable of sustaining flight.

The naturalists of the day soon improved on Pigafetti's account. In 1555 a German zoologist, Konrad Gesner, asserted that the back of the male had a hollow in which the female laid her eggs in flight. She sat on the eggs while the male floated on in the marital flight. Another early naturalist wrote: "Their true home is the terrestial Paradise where their sole sustenance is the dew of heaven." There was often more fantasy than fact in early natural history.

Even a voyage to the Moluccas toward the end of the sixteenth century didn't prevent Johannes Huygen van Lin-schoten from writing this erroneous but delightful nonsense: "In these Islands onlie is found the bird which the Portingles call passeros de sol, that is Fowle of the Sunne, the Italians call it Manu codiatas and the Latinists, Paradiseas, and by us called Paradice-birdes, for ye beauty of their feathers which passe al other birdes; these birdes are never seene alive, but being dead they fall on the Ilands: they flie . . . alwaies into the sunne, and keep themselves continually in the ayre, without lighting on the earth for they have neither feet nor wings, but onlie head and body and for the most part tayle."

Thousands of specimens were imported to Europe in the decades that followed. But, oddly, three centuries were to pass before the first live birds were shipped to Europe—to shatter the romantic legend of a wingless, legless bird that supped on honeydew. The first live pair was brought to the London Zoo by the famous zoologist and collector Alfred Russel Wallace in 1862. (Wallace was coauthor with Charles Darwin of the *Origin of Species*.)

Afterward, Wallace must have wondered about the consequences of his innocent action. If the birds of paradise had been a sensation in 1522 when they had delighted the king of Spain, in 1862 they were a furore! Fashionable ladies all had to have plumes of these lovely birds, and the world's zoos all wanted them.

Hunters plunged into the jungles of New Guinea and its adjacent islands, the main habitat of the birds. Naturalists, too, were deep in the jungles, seeking new species. Some they found were named after European royalty; one was named for the emperor of Germany, and another after the tragic

Austrian Crown Prince Rudolph of the Mayerling tragedy. This last was a superb bird with vivid blue wings and velvet-black plumes, which today is also known as the blue bird of paradise. The name of Rudolph's wife, Princess Stephanie, was bestowed on a bird with long, gleaming black tail feathers.

Birds of paradise in their natural habitat had maintained their numbers against natural enemies and Malay hunters. But European rifles almost wiped them out. We shall never know the full extent of the carnage during the latter half of the nineteenth century, but in 1895 750,000 plumes were sold from *one* London warehouse. Less than 20 years later, sales had dwindled because of the slaughter; only 30,000 were sold in 1913 in London, the world's clearinghouse for the plumes.

Naturalists, meanwhile, had been agitating for protection of the birds, and the United States led the way by banning the import of plumes in 1913. Shortly after, Mexico followed the American lead, and Great Britain in 1921. The world's loveliest birds were thus given a breathing space.

Now, sadly, they face a fresh threat. At a recent "sing-sing" (festival) in the New Guinea highlands, some 2000 dancing natives wore about $200,000 worth of princely plumes. Most of the dark-skinned Melanesians wore at least a pair, many four to six plumes, and a few tycoons up to sixteen! The favored style was the ends of two plumes pushed into the septum of the nose, so that the plumes curled up and framed the face. The more opulent natives wore others as top-lip adornments and as headdresses.

That $200,000 figure is conservatively based on the prices prevailing before 1913, when a single plume fetched $40. Recent sales in the East have reputedly been for as much as $100 and even $200 for the now rare plumes.

In a somewhat sour moment a naturalist once dismissed birds of paradise as "glorified crows." Some crows! Birds of paradise are only very distant relatives of the crows; they are a distinct and diversified family in their own right—Paradisaeidae. Most are small; they range in size from that of a thrush to that of a dove. In the widest sense, including the riflebirds and manucodes—the original birds the Malays had gathered—some 60 species are now recognized. (The name "manucode" is derived from the Malay *manuk dewata*.) They make up six subfamilies: riflebirds and their allies; Long-Tailed birds of paradise; sicklebills; king birds of paradise; typical birds of paradise; and the manucodes and paradise crows. About 40 are found only in New Guinea and adjacent islands.

But peace and prosperity in New Guinea have brought no blessings for the loveliest birds in the world—only death and the threat of extinction. Before the Australian government brought peace to New Guinea, the quick-tempered highland natives were too busy hunting each other's heads to spend too much time on lavish "sing-sings." Now they have good wages from white settlers and money from their own cash crops. The Melanesian Joneses are buying plumes from natives in other areas. All over New Guinea the birds of paradise are being hunted.

Settlement, too, is a threat. Many of the species can't live higher than 6000 feet, but steel has already bitten into trees up to 8000 feet.

Some years ago the Australian government banned trading in plumes among natives, while still permittimg them to hunt with bows and arrows. A more positive step would be the setting aside of large virgin areas as sanctuaries for these dazzlingly lovely birds—and other fauna and flora. The proposal is under consideration.

Meanwhile, as a step toward the preservation of the deservedly named birds of paradise, the Australian millionaire philanthropist, the late Sir Edward Hallstrom, established his own sanctuary at Nondugl in the New Guinea highlands. Sir Edward also gave generous collections of living birds to the world's zoos, including a princely gift in 1953 of 80 birds worth $170,000 to American zoos.

Some 60 species of manakins make up the Pipridae family of Central and South America. The males clear an elliptical court about 30 inches long by 20 inches wide on the forest floor, meticulously removing every leaf and twig. The courts are usually about 20 or 30 feet apart—though in some species they almost impinge on each other—and they usually follow the line of a lake or stream. In dense jungle, visibility is restricted, but the male Gould's manakin (*Manacus vitellinus*) makes himself known to the females by producing a loud snapping sound with his specially thickened wing quills. The resonant snap of his wings can be heard up to a quarter of a mile away by our ears—and presumably farther by the more acute ears of female manakins. When a female approaches, all nearby males begin flashing through the air and calling. Nearing his court, a male enters it with a loud snap of his wings and then leaps from side to side, accompanying each jump with a snapping sound. If a male is successful in his courtship, the female flies into his court and the two birds leap across the court, passing each other in midair. The courtship completed, they fly off into the jungle to mate.

Nonvocal noises, produced by the wings, are used by a number of male manakins to attract females—a kind of bird "wolf whistle." These dramatic sounds have been likened by listeners to the muffled rattling of dried peas in a pod and to the crack of a whip, and described as "a sharp explosive snap" and a "sharp percussive crack."

Even more eye-catching are the whirling dances of the yellow-thighed manakins (*Pipra mentalis*). A number of males compete with each other by turning themselves into wheels of color, each on his own branch, whenever a female approaches. The "whirling dervish" display, as Professor Joel Carl Welty aptly calls it in *A Life of Birds*, starts with the yellow-thighed manakin stretching high on his legs to display his brilliant yellow thighs. Then he tilts his black body forward at a right angle and bends his brilliant scarlet head downward. Then, with wings partly open, the bird swings round the limb, facing first forward and then backward, so that his scarlet head blurs into a ring of color. Another courtship performance of these birds is for each male to slither backward along a limb for about a foot or so and then jump back to where he started and repeat this stiff-legged dance over and over again. The male accompanies this with loud snapping sounds, presumably produced by his wings.

So wonderful, indeed, are the courtship ballets of the manakins that anything is possible. There have been reports of circles of males hopping up and down in time to a song that another manakin outside the circle was singing! What is vouched for are the extraordinary accounts of the dances performed by two long-tailed manakins (*Chiroxiphia linearis*). The males have brilliant red caps, sky-blue backs, long, curved tails and bright-orange legs. In one dance two males perch on a branch near the ground with their heads both pointing in the same direction along the branch. In turn each male rises straight in the air for about two feet, uttering a guttural catlike mewing sound, and then descends to the spot from which he rose. Thus the two birds rise and descend, rather like brightly colored pistons! As the intensity of the dance rises, the leaps become swifter and shorter and the birds become frenzied. On other occasions the two birds perform a dance in which one observer* said, "The two birds replace one another with cyclic

---

* Slud, "The song and dance of the Long-tailed Manakin, *Chiroxiphea linearis*," *Auk* magazine, 1957.

regularity . . . like balls in a juggling act." It begins with the two males facing in the same direction along the same branch. The front bird rises into the air, crying like a cat, and hovers about two feet in the air. The other bird slides forward along the branch until he is underneath the hovering bird, which flies backward in a parabola and alights on the spot which has been left by the other bird. Then the second bird, who is now in front, rises into the air; the other bird shuffles along the branch until he is underneath the hovering bird. The brilliant dance continues, speeding up until the two birds seem almost to make a wheel of brilliant color.

Lek birds are the Carusos and Nureyevs of the bird world. In the next chapter we'll look at other highly talented lek birds.

32

MALE DANCERS AND SINGERS: PART TWO

GROUSE AND MANAKINS, as we have seen, are ballet dancers and vocalists. The bowerbirds (Ptilono-rhynchidae) of Australia and New Guinea are, too—and also architects, gardeners and painters. Bowerbirds are also talented mimics who can imitate the calls of other birds with high fidelity.

We can imagine the astonishment of the famous Italian botanist Otoardo Beccari who in 1878 sent back a report from New Guinea of a wonderful bird living in jungle so thick "that scarcely a ray of sunshine penetrated," and which had built a "conical hut or bower close to a small meadow enamelled with flowers." The meadow, Beccari said, was artificially made of transplanted green moss and had freshly picked flowers, fruit and fungi of vivid colors strewn over it. He added that the bird that had planted this garden had built a sort of thatched hut around a slender sapling, and round the base it had attached a cone of plant fibers.

There were even greater wonders to be revealed over the years to wondering naturalists as they studied the 16 species of bowerbirds—seven in Australia, eight in New Guinea and one shared.

Some bowerbirds build platforms; others build maypoles; and others yet build avenues. Maypole builders are birds of the rain forests; for their maypole they use a sapling and pack a cone of fabric around its base. The golden bowerbird (*Prionodura newtoniana*), which is only nine and a half inches long, may build a nine-foot-high cone of fabric. The golden bowerbird of north-eastern Queensland, Australia, is sometimes called Newton's bowerbird, to honour the naturalist Professor Alfred Newton; and sometimes the golden gardener bowerbird because of its habit of planting seedpods, lichens, moss and orchids on its tall maypole. Besides this tall wall-like structure, the male frequently builds another shorter wall around a sapling growing within a few feet of the first one. A feature of the bower is a vine or piece of wood that extends between the two trees and forms a special "bridge" or display stick where the lovely male bird can display his golden fantail, his small erectile crest and wings and do his decorating.

Another maypole builder, the brown bowerbird (*Ambylornis inornata*) of New Guinea builds a conical waterproof hut above a dwarf cone. The open-sided structure looks very much like the thatched hut Beccari described. A feature is the incorporation of living orchids into the walls of the hut. Another maypole builder surrounds the bower with a low parapet of moss, which it carries to the site. Yet another bowerbird who builds a conical hut surrounds the dancing area with a low stockade of twigs and grass. The bases of the maypoles are usually decorated by the bird with flowers, berries, leaves and beetles' wings of various colors. Each species of maypole builder favors different things for its decorations. In these "theaters" the males display with song-and-dance routines.

The first accounts of the bowers and their decorations

caught the imagination of the Victorian public. Charles Darwin, when he visited Australia, was fascinated by these apparently artistic birds, and so was Thomas Henry Huxley, who later also visited. John Gould, the famous English ornithologist, is believed to have coined the name "bowerbird," which became a household term applied to people, and especially to little girls, who collected odds and ends of colorful rubbish without discrimination. Indeed, a bewildering amount of bric-a-brac was found in the bowers of the avenue builders in Australia—not only shells, pebbles, bones, precious opal, quartz, but teaspoons, gold and silver coins, broken glass, nails, beer-bottle tops and brass cartridge cases. Even false teeth and gold-rimmed spectacles were found in some bowers. Most of these birds, however, are extremely discriminating. For instance, the satin bowerbird (*Ptilonorhynchus violaceus*) of eastern Australia, a beautiful iridescent blue-black bird about the size of a small pigeon, has a pronounced taste for blue objects. He builds a bower with two parallel walls of twigs and decorates it with blue objects. His passion for blue is so strong that he has often stolen bags of bluing from outback laundries in Australia. And he will steal blue flowers such as delphiniums, petunias and hyacinths from gardens. Not long ago a naturalist, Frank Snow, wrote to an Australian newspaper that a satin bowerbird's playground, not 150 yards from his front door, had an all-blue collection of more than 40 articles, ranging from a blue tin toy elephant and toy truck to a blue ball-point pen, blue bus tickets and a ladies comb—not to mention some of his finest polyanthus blooms! The late Professor A. J. Marshall found convincing evidence of the birds' passion for blue by an experiment in which he placed many fragments of broken blue glass bottles near known bowers of the bird over an area of about 130 square kilometers (50 square miles). Professor Marshall numbered each piece with a diamond pencil. He found that within seven days 34 percent of the glass fragments were in the bowers, and a month later 79 percent. He checked on the bowers during the next two years and discovered that these numbered bits of glass moved continuously among the bowers. This came about because the male birds continually raided and counterraided each other's bowers. Thus, Professor Marshall discovered, the more aggressive the male, the bigger the collection of these glass jewels he owned.

In another experiment* Mr. Gaston C. Renard placed a number of colored cards before the bower of a satin bowerbird. Each card, about two and a half by one inches, had a single color—blue, purple, red, green, orange, yellow, brown, white, and black. Early the next morning the naturalist took up his position in a well-concealed spot a few yards from the bower. He had not to wait long before the male arrived. The bird saw the colored cards almost immediately, investigated them and gave voice to a throaty hiss. He eyed them briefly, then seized the red card and, hissing, carried it about 14 feet away and dumped it. (The bird has several cries—the most common is a loud "whee-ooo.") He then returned to the bower, picked up the blue piece and placed it on the platform at the entrance to his bower. He did the same with the purple piece. Then he saw the orange card, pounced on it noisily and carried it away from the bower. In turn he carried away the white, green and yellow pieces. The brown and black pieces were ignored. All the time the bird was uttering his characteristic whee-ooo-ing cries. Later the same day the naturalist repeated the experiment at a bower of another bird and achieved very similar results. Later he could not even *find* the red piece, which the bird apparently disliked intensely, although he searched within a radius of 20 yards of the bower. The only difference was that this bird had apparently accepted the yellow card, which now lay on the platform.

---

* Described in *The Australian Museum* magazine, January–March, 1946.

In another experiment Mr. Renard prepared six cards of about the same size as the others. He colored half of each card blue and the other half respectively brown, yellow, red, white, black, and orange. The cards were painted blue and another color on both sides. When the bird discovered them, he examined each for some minutes, picking them up and turning them over. After this apparent puzzlement, he placed the blue-and-black card on his platform; then he picked up the blue-yellow card and deposited it there. In quick succession he picked up and placed the remainder—blue-red, blue-orange, blue-white, and blue-brown—on his platform. This little experiment, Mr. Renard suggests, showed that the bird's preference for blue was apparently sufficiently strong to overcome its dislike of other colors such as red.

Adult satin-bowerbird cocks are territory owners even when they are away. Young cocks will often approach unoccupied bowers but will rarely dare to enter. But, emulating their father or uncle, they sometimes bring a blue feather or flower to the bower. When the owner arrives, the young males are routed with vigorous scolding and pecking. Entering the bower, the owner appears to indulge in some strenuous housekeeping. He picks up a leaf and puts it down, or turns over a blue flower and arranges it somewhere else, or straightens a twig or two, all the while singing to himself and mingling with his own calls brilliant imitations of the calls of other birds in his neighborhood. I've heard a satin bowerbird give perfect imitations of up to six different birds.

Another species, the spotted bowerbird (*Chlamydera maculata*) is an even better mimic; one once deceived me badly for a short time in central Australia. The sparse desert seemed to be alive with birds and birdsong—but I couldn't see the birds. I discovered that the sole musician was a spotted bowerbird. A spotted bowerbird once deceived a Queensland sheepfarmer with two imitations of thunder during a drought when the farmer was anxiously scanning the skies for rain. He rushed inside crying excitedly that rain was coming. He later realized that he had been fooled. Another story of the spotted bowerbird's remarkable memory has been reported by the Australian ornithologist Alec Chisholm. "A woman in Queensland was astonished to hear the voices of her sister and two daughters talking below the balcony of the house, when she thought they were miles away," he writes in *Bird Wonders of Australia*. "All were talking together as was their habit. The mother's voice sounded soothing and explaining and those of the girls rather aggrieved. The sister went outside to see what was happening. The whole performance came from a spotted bowerbird perched in a mulberry tree."

All seven Australian species of bowerbirds seem to specialize in reproducing unusual sounds. They have successfully imitated the barking of dogs, the noise of cattle breaking through scrub, the creaking of branches in a high wind, the clang of a maul striking a splitter's wedge, the shouting of men, the sound of cornet and violin, of sheep walking through fallen dead branches, of emus rushing through twanging fence wires, the squealing of trapped rabbits, the croaking of frogs, the snarl of a circular saw, the cracking of bullwhips, the clatter of galloping horses, the whistling wings of pigeons—all these and more.

The spotted bowerbird seems to prefer pale-colored objects, such as bleached bones, shells, broken glass, bits of tin, thimbles, screws, spoons, forks, coins and the like. (Gold sovereigns and half-sovereigns sometimes ended up in their bowers in Victorian times.) He has been observed to hop through an open window onto a dressing table to steal jewelry, and one ambitious bird stole the keys from a parked car. The owner, unable to prevent the theft, recovered his keys by searching for and finding the bower. Little in fact is safe from bowerbirds if they take a liking to it. One bowerbird stole a glass eye that a country man in Australia had placed on his bed table!

Some bowerbirds paint the inside walls of their bowers. For instance, the satin bowerbird paints the inside of its avenue with the juice of fruit it has chewed to a pulp or sometimes with charcoal. To watch a satin bowerbird use charcoal for painting is to observe a fascinating operation in which the bird uses a tool of his own making. The bird gathers the charcoal and grinds it with his beak to a sticky black paste. Then he picks up a tiny wad of bark with his beak. It used to be supposed that the bird used this piece of bark as a paintbrush, but actually he uses it as a stopper to keep his beak slightly open and allow the charcoal paste to ooze out the sides. He smears the charcoal energetically all over the twigs of the bower and will continue to do so several times a day until it is covered with a thick plaster of sooty black. As an experiment, Renard once placed a small quantity of powdered bluing on a piece of bark before the bower of a satin bowerbird. The bird arrived and according to his custom began to tidy up the bower and freshen up the charcoal paint on the walls of its play hall. When he discovered the powdered bluing, he showed great excitement, particularly when he tried to pick it up and discovered its consistency. He made no attempt to use the powder on this occasion. But the next morning he again appeared to show the greatest interest in the pigment, and after some excited chatter began to smear it, mixed with saliva, on the walls of the play hall. He continued to do this for half an hour before abruptly leaving the bower. Over the next three days the bird used every scrap of the bluing and ignored the charcoal. But unfortunately Mr. Renard had to leave the area and was unable to carry the experiment further.

A number of other bowerbirds plaster the inner twigs of their bowers with fruit or grass mixed with saliva.

Such apparently artistic activities have led some naturalists in the past to credit the bowerbird with great intelligence and with an aesthetic sense. But Professor Marshall, who studied bowerbirds for over 20 years, puts it down to the sexual instinct in an article in *Scientific American*. "Bowerbirds seem to be no more intelligent than any other crow-like birds," he says. "They will display fully and build bowers only when sex hormones are flowing in their blood streams. The bower and its decorations are the focal point of the male territory, and to it the male attracts the female. The bird's display, often violent, even frenzied, keeps off rivals and holds the female's attention until the seasonal period of reproduction is due." As Dr. Marshall points out, many male bowerbirds "build a bower out of season." Satin bowerbirds, for example, have been observed to build bowers in every month in the year, although the female is normally only ready to mate in October, the spring season in Australia.

Mating between the bowerbirds takes place near the bower or sometimes in the bower. Thereafter, the female goes off alone to build her modest nest in a tree and later to care for the young. The male continues to retain his interest in the bower; he will continue to display there, drive off intruders and raid other birds for their ornaments.

The prolonged courtship of bowerbirds poses some questions for ornithologists, but Professor Marshall believes that it is advantageous for the male to stake out his territory early. "Consequently long before the arrival of the season for reproduction he begins his noisy, flashing, even violent displays to keep rivals away and hold an eligible female within his sphere of influence, so to speak," he says. "When the Eastern Australian forests become full of a harvest of insect food in the late spring, the female is finally ready to accept the male."

How did bower building start? It is probable that it had its origins as a form of displacement of energy. The males as well as the females of many perching birds possess an inherent urge to build nests. In the male bowerbirds it has been diverted into the building of bowers.

Another fascinating question that interests ornithologists is why particular bowerbirds show passions for particular colors. The clue may lie in the fact that the male birds appear to prefer their own colors. Thus, the blue-black satin bower-bird has a penchant for blue objects. It could be that the objects remind him of himself—or of other male birds—because he has been observed to worry these blue objects quite fiercely. A romantic suggestion—not to be taken too seriously—is that the male satin bowerbird picks just the right shade to match the light blue of the female bird's eyes. (The female birds, incidentally, are brown with greenish wings.)

Another maypole dancer is Jackson's whydah (*Euplectes jacksoni*) of East Africa. The male dances around a ring of grass he has trampled down around a central tuft. The distinguished American ornithologist Roger Tory Peterson suggests that the central tuft perhaps represents a symbolical nest. Rather like prairie chickens, Jackson's whydahs assemble in the breeding season in arenas in which as many as a hundred males dance around their maypoles for the favors of the females.

Among lek birds, the lyrebirds of eastern Australia are remarkable for having not just one arena but as many as a dozen. Throughout his forest territory the male of the superb lyrebird (*Menura superba*) scratches up a number of display mounds, about a foot high and three feet across. During the breeding season—and even earlier—he visits each of these in turn in order to display and sing. The brown-colored superb lyrebird resembles a small pheasant in appearance—the first settlers called them "native pheasant," "Botany Bay pheasant" and "New South Wales birds of paradise." The resemblances are superficial, for the lyrebirds belong to the song-birds (Oscines) and are of very ancient lineage. Because the birds are shy and wary and their wet valley habitats are thickly overgrown with scrub, they are more often heard than seen, even by experienced naturalists. (Like other birds who use sound to guard territories in thick forests, the superb lyrebird has powerful if musical lungs.)

The spectacle of the male superb lyrebird dancing and singing on the "stage" he has scratched up is one of the most richly rewarding aesthetic experiences in nature that I have been fortunate enough to enjoy. A naturalist friend and I waited at noon on an April day in a hide he had prepared with a good view of a bird's stage in a small clearing. This male bird had six of these stages, which he visited throughout the day. (Some may have up to a dozen.) He was expected here about one o'clock. It was a mild late-autumn day. Tall, slender eucalypts with feathery canopies reached up into the sky. Somewhere below us, hidden by bushes and ferns, was a small stream. The forest floor was heavy with damp leafmold.

At ten minutes to one we caught our first sight of the male. We'd heard him earlier, making the rounds of his stages to perform his song-and-dance routines and feeding on the way. Lyrebirds spend most of their time on the ground, scratching for food with their powerful feet through fallen leaves or tearing decaying logs to pieces for insects, worms and small land mollusks.

The dark-brown male advanced into the clearing. He was about the size of a small domestic fowl, with short, rounded wings. As the sun caught his sides, green-brown and copper tints flashed. His breast and belly were brown-gray merging to light gray. His folded tail, almost trailing on the forest floor, was about two feet long. The two outer plumes were broad and curved like the arms of a lyre; they were white, orange and lead-colored, with black tips. We couldn't see all of the 14 inner filmy plumes and wouldn't unless the bird decided to display.

"The male starts to display even before the winter breeding season from May to September," my friend had explained earlier. "He'll probably display today without the female—she isn't quite ready to mate."

The male stalked cautiously into the clearing, cocking his head from side to side, listening intently. His large brown eyes ranged about. Reassured, he raked at a patch of leafmold. Three thrusts with his thick right leg, four with the left. The long claws dug deep. The small head darted and plucked out first a worm and then a centipede. He swallowed them with an upward flick of his head, then went on feeding. Food was plentiful; no wonder his ancestors had long ago taken to an abundant life on the forest floor, only flying when disturbed, with quick, hurried wingbeats.

We waited tensely, not moving. Put a foot on a dry twig, make a sudden movement that caught his large protruding eyes, and he'd dart away. We searched with our eyes. No female. Would he display or go on feeding?

He caught sight of his stage and ran forward. He stood on the mound and began to sing, softly at first. His tail plumes rose until they were vertical, now spread wide into the full shape of a lyre. He lowered them forward over his back like an opened fan, to touch the ground in front of his head and cloak his iridescent, coppery, glinting body with a shimmering, silvery, diaphanous mantle. The inner plumes were fine as spun glass with the filaments gleaming in the sun like fine steel wires. He sang and danced ecstatically, advancing and retreating, sometimes circling, always in time to the cascades of sound from his pulsing throat. His lovely gossamer cloak shimmered. This was his own song, no mimicry as yet, but a glorious blend of his own calls, spanning over three octaves. (One talented lyrebird in Sherbrooke Forest, near Melbourne, Australia, covered four.) He sang on, tremulous with excitement, and then he began to mimic the calls of other birds, of currawongs, kookaburras, gray thrushes, blackbirds, scrub wrens, pilot birds, bellbirds. The mimicking transcended mere imitation; the bird was an artist and improvised variations on the calls, weaving them into a pleasing tone poem. His recital—bolder, more exhilarating than any nightingale's—lasted eight minutes while we watched spellbound.

Recitals sometimes last for 20 minutes—and nearly all day if the bird has a nearby male rival.

The other Australian lyrebird, the Prince Albert lyrebird (*Menura alberti*), of the rain forests of southern Queensland and northern New South Wales, does not build stages. It is suggested this is because its tail is not as spectacular as that of its larger cousin. However, it displays throughout its territory both on the ground and on logs.

Both species breed from May to September, and the females of both lay usually a single egg varying in color from light stone-gray to deep purplish-brown, with streaks and spots of deep slate-gray and blackish-brown distributed over the surface. Like the bowerbirds and the American mockingbirds, both males and females of the two species are superb mimics. Indeed, some ornithologists, such as the leading American expert Professor Charles Hartshorne, rate them as the best of all. Some of the well-vouched-for accounts of the lyrebirds' skill in mimicry are almost unbelievable. An Australian writer took his typewriter into the forest country in order to finish a novel. After some weeks he was unnerved—particularly at times when inspiration flagged—to hear "typewriters" clattering all around him in the bush. The hi-fi reproduction was coming from a lyrebird or lyrebirds. Then there's another astonishing story of the superb lyrebirds' deceitful skill, which is recorded in *Bird Wonders of Australia* by the Australian ornithologist Alec Chisholm, who knows more about lyrebirds than anyone living. It seems that timber workers at a timber mill in a Victorian eucalypt forest had an agreed code: Three blasts of a whistle would tell them of an accident, six of a fatality. One day six blasts echoed and reechoed round the hills. Men ran to the mill, but there had been no fatal accident; the six blasts had been given by a superb lyrebird. It had heard the three blasts for an accident fairly often and had imitated them, adding three more for good measure!

Lyrebirds have also faithfully mimicked the heavy sounds of axe blows thudding into tree trunks, the twanging of wire fences (presumably caused by kangaroos, sheep or other animals), the noise of car horns and engines, the puffing of trains, etc. One superb lyrebird named Jack was a pet on a farm in Victoria for 20 years. Jack was said to have reproduced certain words and phrases and several bars of piano and violin music. Another in Victoria is said to have reproduced daily and faithfully the loud noises of a quarry at full blast. Listeners also claimed they could detect occasional words in the bird's reproduction of the men's voices. It sounds like a tall story, but the superb lyrebird's hi-fi talent is extraordinary.

The feats recounted above are the more unusual ones of these forest-dwelling, ground-feeding birds. More frequently the male lyrebirds of both species mimic the calls and songs of other birds in their forest habitats, as did the bird my friend and I observed.

Why do lyrebirds, bowerbirds and mockingbirds mimic? You can always start a lively argument among experts with this question. Many theories are advanced, but no single one explains all instances. However, some things are reasonably clear: Most mimicry is by males, and most talented imitators have some things in common, such as being, for the most part, plain-colored birds living in densely covered areas where vision is restricted and sound is doubly important in maintaining territories. In dense scrub and forest land, bright feathers aren't much use in attracting females or warning off intruding males. Hence, some experts argue, the virtuoso mimics are those birds who have had to develop an extraordinary sensibility to sound.

No one knows if lyrebirds, bowerbirds or mockingbirds gain any advantage from their talent for mimicry. Brilliant mimicry could possibly help its owner by giving him a complex and varied song for advertising purposes; it could make him distinctive and establish his "brand image."

But though mimicry probably has utilitarian reasons, it transcends them. When you hear a lyrebird exuberantly weaving his own calls and songs and those of other birds into a harmonious whole, you are listening to a creative musician. The birds seem to love sounds, to have a lively interest in them and to use them tastefully. The lyrebird—and some other male songbirds—are on the border of art, of self-expression. If not genuine musicians, the Australian lyrebirds are at least minstrels. Some "composing" skill appears to be possessed by superb lyrebirds in Sherbrooke Forest, Victoria, as illustrated by the one I observed, who sang what might be called variations, adapting and changing the calls. It is probably significant that lyrebirds and some other songbirds are responsive to our own music. Thus, the European nightingale can often be started into song by the playing of a violin. An English amateur naturalist has claimed that he had a pet blackbird that clearly derived more pleasure from Mozart than any other composer.

To conclude, there are good reasons why songbirds are highly skilled musicians or, at the very least, minstrels. Sound is the biggest thing in their lives, as sight is in ours. There is also a physiological reason why birds should be as highly emotional as they clearly are. They live more passionately than we do because their metabolism is much higher; as we saw earlier, their body temperatures are up to 12 degrees higher; their pulses beat faster and, in the words of Sir Julian

Huxley, they live in what would be fever heat in ourselves.

While birdsong may have its origins in utility, as we have suggested earlier, it appears to transcend it. It becomes the means by which the male seems to express his strong emotions of ardent love, triumph and rivalry. Certainly, if we saw a man behaving in the same way, we would say that he was expressing these emotions!

33

NEST BUILDING

MANY ANIMALS BUILD NESTS, including insects, fishes, amphibians, reptiles, mammals, but none so efficiently as do birds. The popular idea of a nest as a bird's permanent residence is most inaccurate, and references to little birds flying back to their nests at night or seeking the safety of their nests are false and sentimental. Apart from a few species of birds that live in holes in trees, no birds use their nests as permanent homes. Nests are built for the rearing of families. Birds build them to protect themselves, their eggs and their young from bad weather during the breeding season and from predators.

Some birds, such as vireos and orioles of North America seek security by building their nests at the tips of tree branches or suspended from slender limbs; others build in cliff edges or in holes and on or over water.

Many tropical birds weave delicate nests, which they suspend over water. Very often these have a socklike entrance dangling from the nest, which provides some obstacle to snakes

and monkeys seeking the eggs or young. The oropendolas of tropical America are some that do this. Auks (Alcidae) and other seabirds frequently nest on cliff ledges where the eggs are safe from most predators. Grebes build floating nests of debris on water to provide some security for eggs and young; the nests are tethered to reeds or saplings. The British dipper (*Cinclus cinclus*) often builds its spherical nest behind waterfalls. So does a swift (*Collocalia gigas*) of Java, which arrows through mist and spray on every visit to its nest. The ultimate in water protection is that used by the cave swiftlets (*Collocalia francica*) of the Andaman Islands in the Indian Ocean. Large colonies of these birds nest in caves whose entrances are completely submerged when each wave is at its peak. The birds swing round the cliffs waiting for the waves to recede. When the cave entrance is uncovered, the birds swoop into the mouth and then up to their nests on the walls above sea level. Some cave swiftlets of the Himalayas breed deep down in potholes and in underground caves. In one hole 240 feet deep, the nests started 50 feet below ground level and covered both sides to the very bottom.

Camouflage, too, affords some protection of nests. As you've probably discovered, some birds' nests are very difficult to find. Here natural economy of effort plays its part. Thus, birds who nest in grasslands tend to use grass for their nests, and this blends with the habitat. So, too, with birds who use reeds and build among reeds. Some birds, such as hummingbirds, cover their nests with lichens, which helps to merge them into the background. Some ground-nesting birds that build no nests at all probably carry camouflage as far as it can be carried. The most notable of these are some plovers and goatsuckers whose eggs so much resemble the terrain that they are extraordinarily difficult to find.

Some birds may seek some security by building communal nests. A hundred pairs of social weavers (*Philetairus socius*) will pool their labor to build a large, flat dome of grasses. This is usually built in solitary trees on the African veldt, and very large ones often resemble a native hut. The dome completed, each pair builds its own nest on the underside with narrow, tubular entrances. One incomplete structure—part had been broken off—was 25 feet by 15 feet by 5 feet to 10 feet in height and had 95 nests.

Other birds seek security by building nests close to larger birds, frequently birds of prey; small perching birds, such as sparrows, grackles and weaverbirds, have been known to build their nests on the edges or underneath nests built by eagles, hawks, ospreys and owls. They have also built nests under those of storks. Some African weaverbirds build near the nests of jackal or augur buzzards (*Buteo rufofuscus*). Cliff swallows (*Petrochelidon pyrrhonota*) often build near the nest of the prairie falcon (*Falco mexicanus*). This proximity to birds of prey may look hazardous, but usually the predators seem to take little notice of their tenants. On the average it must pay, even if the bird of prey occasionally takes one of the smaller birds.

Other birds achieve safety not in community nests but in nesting in large colonies, notably seabirds such as penguins, petrels, gannets, pelicans, gulls, terns and auks. (Some rookeries of Adélie penguins in Antarctica are estimated to have one million pairs of breeding birds.)

Some other birds seek security by building their nests near the nests of stinging insects, such as ants, wasps and, sometimes, bees. In Australia the black-throated warbler (*Gerygone palpebrosa*) nests so frequently near hornets' nests that it is called the "hornet-nest bird." In South America, sometimes up to ten nests of the yellow-rumped cacique (*Cacicus cela*) hang from trees around the nest of some particularly aggressive wasps. Other birds even make their nests *inside* the nests of stinging or biting insects. For instance, the buff-spotted woodpecker (*Campethera nivosa*) of Africa regularly builds its nest in the middle of the round, papery nest of tree ants. Some parrots in New Guinea and Brazil make their nests in those of tree termites, excavating them with their claws. A number of kingfishers and parrots in Australia also excavate their nest burrows in the nests of termites.

What is remarkable about these curious partnerships is that the insects do not trouble the birds or their young. The birds, on the other hand, are not only protected but sometimes eat the insects and their larvae.

One of the most interesting associations of this kind is that of the village weaver (*Ploceus cucullatus*), which was introduced from Africa to Haiti about 300 years ago and now commonly builds its nest there near stinging wasps. In its African home, however, it habitually nests with predatory birds or with man.

Some birds associate with animals, but not necessarily for protection. For instance, the minera (*Geositta cunicularia*), a relative of the ovenbirds of South America, nests regularly in the burrows of a rodent, the vizcacha.

In New Zealand two shearwaters, *Puffinus carneipes* and *P. bulleri*, associate with tuataras, as observed earlier. But here it seems the tuatara chooses the bird's burrow rather than the other way around.

Associations exist between man and a number of birds, such as some species of pigeons, martins, swifts, weavers, house sparrows and common starlings. In some cases our buildings only offer convenient sites for nests. But it is at least arguable that proximity to man also provides protection for birds such as house sparrows and common starlings, which have thrived on the protection of our houses and the scraps of food we leave around. In North America, chimney swifts (*Chaetura pelagica*) probably build more frequently down our chimneys than they do down the hollow trees that were once their usual nesting sites. In the United Kingdom the great tit (*Parus major*) frequently builds its nest deep down the tubular metal posts employed in clotheslines.

In Europe for many years some species of storks, such as

the white and abdim's storks, have bred in the middle of towns and cities. Presumably, the birds find some security here from birds of prey, even if they may have to fly farther to find food for their fledglings.

Some birds, too, have benefited from association not with man but with his domestic animals. In the heyday of the horse, the chipping sparrow (*Spizella passerina*) gave up using fine rootlets for nest building and substituted horsehair. Since the decline of the horse, the birds are reverting by necessity to rootlets.

Among the stranger associations with man are birds that successfully raise their families on moving ferryboats. They include tree swallows (*Iridoprocne bicolor*), barn swallows (*Hirundo rustica*), American robins, and phoebes (*Sayornis phoebe*). A pair of European robins raised a brood in a railway truck that traveled 200 miles shortly after the birds hatched. One parent accompanied the brood as it traveled, feeding them along the route.

Birds sometimes build in what must strike us as very odd places. European robins have built under the hood of a car and inside saucepans.

Thus, the nests that birds build are almost as diverse as the birds themselves—or, as we shall see later, the eggs they lay.

In nest size the range is immense—from the gigantic nests of mound builders and eagles to the tiny nests of hummingbirds. The large nests of eagles are the accumulation of many years of building on the one site. Generations of eagles have bred in the same nests for up to 60 years. One celebrated nest of a bald eagle (*Haliaeetus leucocephalus*) at Vermilion, Ohio, was 8½ feet in diameter, 12 feet deep and weighed about 2 tons. European sea eagles (*Haliaeetus albicilla*) also build immense nests. One nest that crashed during a storm weighed 2 tons.

Storks, too, build large, heavy nests. Those of white storks have been found to be 5½ feet in diameter and 6½ feet deep. Probably the largest nests of all are built by the hen-size male of a mated pair of megapodes; over the years the mound of a mallee fowl (*Leipoa ocellata*) of the arid regions of southeastern Australia may grow as high as 14 feet and be 35 feet in diameter. The nest of a near relative, the jungle or scrub fowl (*Megapodius freycinet*) can be as high as 20 feet and as wide as 50 feet. These remarkable birds use the heat generated by decaying vegetable matter to hatch their eggs. (So much else about them is also remarkable that we will look at them more closely in a later chapter devoted to mound builders.)

At the other extreme are the tiny nests of hummingbirds. The smallest known are those of the Trochilidae, whose nests may be ⅘ of an inch across and 1⅕ inches deep.

There is also considerable diversity as to who builds the nest. Among some species both share more or less in the building. These include kingfishers, woodpeckers, some swallows (*Riparia*) and waxwings.

Nest building is shared, too, among male and female winter wrens (*Troglodytes troglodytes*). (This species, sometimes called the "European wren," is called the "winter wren" in North America.) Often the impatient male builds several dummy or "cock" nests before the female arrives in his territory. His bride may then select one and line it with feathers and wool.

Among other species the male alone builds the nest. These include, besides the mallee fowl, some shrikes and the Philippine weaverbird (*Ploceus philippinus*). In other species the female builds without any help from the male. These include hummingbirds, manakins, and the red-eyed vireo (*Vireo olivaceus*). The male frigate birds build with material provided by the females. Among ravens and rooks, the female builds, but both sexes gather the material.

Many nests are complicated and intricate structures. Yet

they must all have had simple origins just as sometime in the remote past man built his first crude shelter. One theory, mentioned earlier, was that nest building began from birds directing aggressive feelings during mating toward tearing up grass stems. Other ornithologists believe that nest building started from movements of female birds during mating. The English expert E. A. Armstrong cites the instance of the female tern which pivots on her breast to face the male, bearing a fish gift while he circles about her on the ground during the courtship; in time she makes a saucerlike depression in the sand. She may at the same time pick up pieces of grass and twigs and drop them near her. Mr. Armstrong points out that some species of terns lay the egg in this depression without further ado. Other species of terns add pebbles or sticks. Eggs laid in this simple depression would be better protected than those merely laid on the sand. Another theory is that an ancestral bird accidentally laid its egg on debris washed up on the shore rather than laying it on the bare sand. As a consequence the eggs were better protected than other eggs and more hatched. The offspring remembered the stick "nest" they saw when they were hatched, and sought the same environment when it was their own turn to lay eggs; the first evolutionary step had been taken.

From simple beginnings the process of nest building has become elaborated over the ages. Each species of bird builds its own particular kind of nest. The American robin, for instance, does not build a nest like that of the magpie (*Pica pica*) of Eurasia and North America—and vice versa.

Intricate as some nests are, building is innate; the bird simply responds to one impulse after another—there is no conscious thought or planning. No bird has to learn how to build its nest. It inherits its skill—or lack of it in some species—from its ancestors. For instance, four generations of caged weaverbirds were given no chance to build their normal complicated nests, but the fifth generation built them as expertly as ever when supplied with the right materials.

The materials used are also diverse. In human terms you could call birds carpenters, drillers, weavers, tunnelers, tailors, plasterers and general laborers.

Among the finest of bird artisans is one of the weaverbirds of Africa, who can tie knots of a simple kind. The male birds build hanging nests; such a nest must be built with one or several long grass strands suspended from a branch. The red-billed weaverbird or red-billed quelea (*Quelea quelea*) holds one end of a grass strand under his foot, then winds it several times around the branch and over the held-down strand. Then he ties a knot in the grass strand by tugging the free end through a loop of the fiber he has made around the branch. The cock suspends a number of these grass fibers and then weaves a complete ring of grass. Then he stands in the ring and gradually weaves in other long strands of strong grass until he is almost completely enclosed in a hollow grass ball just a little larger than himself. The final step is the construction of a small vertical entrance tunnel or sock which hangs from the bottom of the nest. (This is a precaution to deter snakes and monkeys.) A necessary refinement is that the bird builds an inner partition to prevent eggs falling into the sock.

Equally skilled are the tailorbirds (*Orthotomus* spp.) of Southeast Asia. The female bores a hole with her beak in the edges of large leaves of tropical plants and then stitches two leaves together on both outside edges to form a green funnel-shaped container for her nest. Using her beak as a "needle," she pulls plant fibers through the opposing sets of holds. She even manages to knot the separate strands together on the outside so they will hold. Alternatively, the hen may stitch together the outside edges of a large hanging leaf to make a purselike nest.

The best-known of the plasterers are swallows. These artisans have to work very hard. To construct a single nest, two barn swallows made over 1200 flights, each time with a tiny pellet of mud. The rufous ovenbird (*Furnarius rufus*) of South America also labors hard. Its ball-like nest of mud and cow dung may contain as much as nine pounds of material—50 times the weight of its tiny three-ounce builder. This mud hut, large enough inside to contain an eagle, has two chambers—an entrance vestibule and a nesting chamber with a half-partition. The ovenbird gets its popular name, of course, because its round nest resembles a baker's oven. In the Argentine, where it is the national bird, the rufous ovenbird is known popularly as *el Hornero*, "the baker." Along many roads nearly every fence post or crossbar on telephone poles has its rock-hard nest of "the baker." The stout little female lays usually five white eggs in the grass-lined inner chamber. Both sexes share incubation. The abandoned ovens are often taken over by cowbirds, wrens, swallows, or parrots.

The ovenbird's technique of building with mud and dung has saved thousands of lives in South America. Some years ago Dr. Mario Pinotti, the head of Brazil's National Department of Endemic Diseases, was wondering what he might do to combat the ravages of Chagas' disease, a trypanosome blood infection that afflicts millions of people in South America. Adults suffer severe debility, and children often die from this incurable disease, which is transmitted by the bite of an insect, the barbeiro, which breeds prolifically in the cracks of native mud huts. Pondering the problem, Dr. Pinotti decided that one way of tackling the disease would be to build crack-free houses—and then remembered that when he was a boy he often threw bricks at the nests of ovenbirds and they never cracked. At Dr. Pinotti's instigation, research workers analyzed the ovenbird's mortar of sand and dung. In a test, 2000 huts were plastered with the birds' odorless cow-dung mortar.

Six months later every house was found to be free of cracks—and of the barbeiro, whereas previously 98 percent of houses had been infested. The work started with the plastering of 200,000 houses in 1958, and since then work has proceeded steadily. It is hoped that within a generation 2.5 million huts will be plastered with the mortar the ovenbird discovered millions of years ago, and that the inroads of Chagas' disease will be severely restricted.

Flamingos also use mud to build their nests. Breeding in colonies, sometimes in enormous numbers, they build truncated cones of mud up to 18 inches high. The birds scoop out a shallow depression on top of the cones in which to lay their eggs. In Australia the magpie lark (*Grallina cyanoleuca*) and apostle bird (*Struthidea cinerea*) build symmetrical bowls of mud on horizontal branches of trees. They reinforce the mud with horsehair and wool. This is an interesting example of birds adapting themselves to changing circumstances, because both horses and sheep are not natural to Australia but were introduced after settlement. The yellow-nosed albatross (*Diomedea chlororhynchos*) builds a nest not unlike that of the flamingos, about 12 inches high, and mixes vegetation with the mud for added strength.

Spider web is put to good use by hummingbirds, titmice, white-eyes, flower-peckers and a few other birds who use it either to hold materials together or to suspend the nest.

Other birds use their saliva as a cement to hold together the nesting materials of twigs, grass, feathers and other matter. They may also use the saliva to glue their hanging nests to tree branches or to the insides of caves. One bird who does this is the chimney swift who cements twigs together to form a small bowl-shaped nest, which it affixes to the inside of a chimney or hollow tree. The swallowtail swifts of tropical America cement saliva and feathers to build socklike tubes up to two feet long, which they stick to cave ledges. Most of the 70 species of swifts use their saliva as a nest cement. The most remarkable use is surely that of the palm swift (*Cypsiurus parvus*) of tropical Africa and Asia. Both birds build the nest. They glue together a simple nest of feathers and plant fluff to make a feltlike pad which they stick onto the inside of a drooping palm frond. The female cements her two eggs, again with saliva, to her tiny nest. Even though the leaf may sway in the wind, the eggs remain firmly attached. The adults are probably the only birds that incubate their eggs in a vertical or near vertical position. Because the birds frequently select a dying leaf, people sometimes make bets as to whether the eggs will hatch before the frond falls!

Best-known of all saliva-built nests are the edible ones of the swiftlets, belonging to the genus *Collocalia*. Two of the species of these cave swiftlets of the Indo-Australasian region build nests entirely of dried salivary mucous. The nests, in the form of small, translucent quarterspheres, are attached to the rock walls of caves. These are the nests that are much valued by Chinese for making the famous bird's-nest soup. Up to 35 million edible nests were exported in one recent year from Borneo for the making of bird's-nest soup. Gathering the nests in the darkness of caves is a dangerous occupation. The collectors use long sets of single poles, pinned or socketed together. Some of these poles may be up to 400 feet high. Men climb the poles and scrape the nests off the wall with little blades attached to the ends of other long bamboo poles.

Hornbills are among the more remarkable of the plasterers. Females wall themselves into their nest holes in trees with a mortar of mud and droppings. (In some species the male helps to imprison his bride!) The female silvery-cheeked hornbill (*Bycanistes brevis*) of East Africa does all the plastering-up of the nest hole. She uses mud of earth and saliva mixed, prepared for her by the male and regurgitated to her in pellets. Like other male hornbills, he brings food to his wife during incubation and sometimes gives her, in addition, flowers! The nest opening left by hornbills is usually so small that the female and her young may have to remain there for several weeks, being fed by the male, until they are able to break out. In some species the female breaks her way out as soon as she has renewed her plumage, and then helps the male with the feeding of the young, who stay in the nest. Amazingly, the young repair the broken entrance, using a mortar of food remnants and droppings. They do this without any help from their parents—surely a remarkable instance of instinct.

Some other birds also nest in holes—using either ready-made ones or in some instances making their own. Kingfishers, motmots and bee-eaters excavate holes in vertical banks. Some birds are prodigious tunnelers; the rhinoceros auklet (*Cerorhinca monocerata*) of British Columbia has on occasion nested at the rear of holes up to 26 feet long. Woodpeckers chisel out their own holes, and usually in sound wood. Less well equipped with nature's chisels in the form of bills, some titmice select pulpy or decaying wood.

Almost anything is nesting material for some birds. For instance, bones are frequently used by owls, penguins, kingfishers, petrels and terns, and over 30 different birds make use of cast-off snake skins. I once saw the nest of an Australian white-backed magpie (*Gymnorhina hypoleuca*) who had used scraps of strong fencing wire to augment its more usual nesting material of twigs and sticks. It is almost unbelievable to think that the bird could have twisted this very strong wire to the circular shape of the nest. Odd materials, too, were used by a Wall Street pigeon who built a nest of paper clips and rubber bands.

Some birds build comparatively primitive nests—notably cormorants, frigate birds, herons, storks, some cuckoos and many pigeons and doves. No one would list these birds among the skilled artisans. Stand underneath a frigate bird's nest and you can frequently see the eggs.

Finally, there are birds who don't build nests at all. These include pratincoles, most auks, nightjars and, most notably, emperor penguins, some cuckoos, cowbirds and honey guides.

The emperor penguin (*Aptenodytes forsteri*) builds no nest for the simple reason that there's nothing to build it with on the sea ice of the Antarctic. Its singular method of incubation in the harshest cold of the Antarctic winter properly belongs in the later chapter on incubation. There is a separate chapter, also, for those parasitic birds—cuckoos, cowbirds and honey guides—that lay their eggs in the nests of other birds (who also incubate them).

Tropic birds (*Phaethon* spp.) dig shallow nests, scraped out with their feet in the debris on cliff edges. Pratincoles, those small relatives of sandpipers, of southern Europe, western Asia and North Africa, make no nest but instead make use of shallow hollows, hoofprints, an unlined scrape or even bare rock. Auks lay their eggs on cliff edges, in crevices, among rocks, under plants or on bare ground and use little or no nesting material. Nightjars usually lay their eggs on the bare ground.

Whatever their diversity of shape and materials, nests are built by their owners to protect them and their eggs and young from enemies and bad weather during the breeding season. In the chapters that follow, we'll look at some of the eggs birds lay.

34

EGGS: RECEPTACLES OF LIFE

I think if required on pain of death to name instantly the most perfect thing in the universe, I should risk my fate on a bird's egg.

T. W. HIGGINSON

THIS NINETEENTH-CENTURY American author and soldier was praising the beauty of coloration of birds' eggs. But there's more to a bird's egg than meets the eye; its size, shape and color may help a particular species in the struggle with other creatures for survival.

Generally speaking, large birds lay large eggs, and small birds lay small eggs. Some gigantic extinct birds laid enormous eggs. The largest egg we know about was laid by the giant flightless elephant bird (*Aepyornis titan*). The 13-by-9½-inch egg of the elephant bird of Malagasy (Madagascar) held over two gallons of liquid, eight times as much as an ostrich egg, the largest laid by living birds. A naturalist broke 19 hen eggs to fill an empty ostrich shell; he'd have to break 100 to 120 to fill an egg of *Aepyornis titan!*

The bird that laid these monster eggs was suitably gigantic, standing nearly ten feet tall. The largest ones may have weighed up to 1000 pounds, as much as a bullock. It looked rather like an ostrich with massive legs. (The legs *had* to be stout to support such an immense body.)

Well-preserved eggs of these giant birds are still being discovered in the bogs of Malagasy. Indeed, the eggs are so plentiful and the shells so strong that people of Malagasy used them as eating bowls. One story goes that the first discovery for science of this heavyweight champion of birds is said to have been made not by a professional zoologist but by thirsty Frenchmen whose eyes popped when they saw the huge eggshells storing rum in a bar. Victor Sganzin, a French naturalist, saw people using half an eggshell as a bowl in 1832. They couldn't be persuaded to part with it, so he made sketches. His story wasn't altogether believed, but not long afterwards large, intact shells were discovered.

Though flightless, the elephant bird was presumably the origin of all the Eastern fables about the gigantic rukh or roc, which was said to carry off elephants to its nest to devour them. You will remember that in *A Thousand and One Nights* Sinbad the sailor was carried by a roc from the island on which he had been left by his companions. One claw of this bird was as large as the trunk of a large tree. On his second voyage Sinbad was left behind by accident on an island. Exploring it, he saw a huge white dome that reached high into the air. He thought it must be a building and walked all around it—a full 50 paces—without finding a door. Then he realized that the great dome was a roc's egg. Other variations of the story exist in the East; the Hindu god Vishnu rides on the back of a giant bird. The legends probably arose in a number of ways. We tell stories ourselves of eagles carrying away babies, which they are unable to do. (In tests an American golden eagle failed to get off the ground with an eight-pound weight.) Elephant birds, which probably persisted into comparatively recent times, could be assumed to lift elephants, given our all-too-human tendency to exaggerate. Or again, the fable may have been based on the eggs, which must have been seen by many travelers. Once started, exaggeration has a natural momentum. Marco Polo told the Great Khan about a giant bird in Madagascar whose wingspan, he said, was 16 paces long (48 feet) and whose wing feathers were 8 paces long. The Great Khan sent an expedition to Madagascar, and it returned with a giant feather. Scholars think it was probably a frond of the raffia palm, which, when dry, superficially resembles a feather.

Eggs of another bird that also became extinct in comparatively recent times, the giant moa (*Dinornis maximus*) of New Zealand, were probably as large as those of the elephant bird. These flightless birds stood nearly ten feet tall, as tall as *Aepyornis titan*, and had equally massive legs. The giant moas were extinct before the Maoris reached New Zealand from the central Pacific, about a thousand years ago, but other, smaller moas existed at the time of the landing and were afterwards hunted to extinction. Well-preserved eggs range in size, according to species, from 7 by $5\frac{1}{2}$ inches to 10 by 7 inches.

There's also an enormous range of size in the eggs laid by living birds—from the pea-size eggs of hummingbirds to the papaw-size eggs of ostriches. In between these extremes is the familiar domestic hen's egg.

Ostriches are the largest living birds and, like the other very large birds, are ratite birds—that is, flightless birds without keels on their breastbones and whose feathers lack stiff vanes and oil glands. A male ostrich (*Struthio camelus*) of South Africa may reach near 8 feet in height and weigh over 300 pounds. The smaller female lays up to 6 to 8 eggs, creamy white and almost spherical—about $6\frac{7}{10}$ by $5\frac{3}{10}$ inches. A typical egg weighs about 49 ounces, just over 3 pounds, and holds about a pint of liquid. For a basis of comparison, the average domestic hen lays an egg weighing about $2\frac{1}{2}$ ounces.

Africans have used the tough shells of ostrich eggs as water containers and eating bowls from time immemorial. They carry bundles of them in netting bags. The archaeologist Sir William Petrie found clay models of ostrich eggs in early tombs in Egypt. Some he found were decorated with an imitation of the network of cords in which they were carried.

At the other extreme is the tiny vervain hummingbird (*Mellisuga minima*) of Jamaica. Its tiny egg is less than four-tenths of an inch long. If you wanted to, you could break nearly 3000 of these minute eggs into an ostrich egg before you filled it.

The next-largest egg is laid by another ratite, the common rhea (*Rhea americana*) of South America who stands about 5 feet high and weighs 44 to 55 pounds. Individual rhea eggs vary considerably, but an average is about $5\frac{1}{2}$ by $3\frac{1}{2}$ inches and weighs about 21 ounces.

Third place in the giant-egg sweepstakes is held by another ratite, the emu (*Dromaius novaehollandiae*) of Australia, the second-largest of all living birds; it stands from 5 to 6 feet tall and weighs up to 120 pounds. The female usually lays more eggs than the rhea, about 8 to 10 (or sometimes up to 16), and they're slightly smaller; the dark-green eggs average $5\frac{1}{4}$ by $3\frac{1}{2}$ inches and weigh about $20\frac{1}{2}$ ounces.

Large eggs, too, are laid by the ratite cassowaries of the rain forests of New Guinea and northern Australia. These birds, with a conspicuous bony helmet or "casque" on the top of their skulls, range through New Guinea, Aru Islands, Ceram, New Britain, Jobi and the eastern coast of northeastern Australia. The light-green eggs of the Australian cassowary (*C. casuarius johnsonii*) vary considerably—from about $4\frac{9}{10}$ by $3\frac{3}{5}$ inches to about $5\frac{3}{5}$ by $3\frac{4}{5}$ inches.

Large eggs are also laid by the largest of the albatrosses, the wandering albatross (*Diomedea exulans*), which is also the largest of all flying seabirds, with a wingspread up to 11 feet, 4 inches and a weight of about 26 pounds. (Exaggerated accounts have persisted about the wingspan of these birds; the Australian Museum at Sydney has been falsely credited with possessing a bird with a wingspan of 16 feet, 4 inches.) These kingly birds of the colder southern oceans, from about 30° south to 60° south, nest on sub-Antarctic islands. The female lays a single white egg, about 5 inches long, weighing about a pound.

Another princely albatross, the royal albatross (*Diomeda epomophora*), lays eggs as large or almost as large as the wandering albatross. Like the wandering albatross, the royal nests mainly on remote, storm-swept sub-Antarctic islands, but about 50 years ago two of them began breeding at Taiaroa Head near Dunedin in the South Island of New Zealand. When I was in Dunedin recently the first thing the conservation-conscious citizens asked me was, "Have you seen our Royal albatrosses?"

To divert briefly, the wandering albatross is probably the one most of us know best because it often follows ships. It was almost certainly a wandering albatross that stirred Herman Melville to this unmatchable description in *Moby Dick*:

> I remember the first albatross I ever saw. It was during a prolonged gale, in waters hard upon the Antarctic seas. From my forenoon watch below, I ascended to the overclouded deck; and there, dashed upon the main hatches, I saw a regal, feathery thing of unspotted whiteness, and with a hooked, Roman bill sublime. At intervals, it arched forth its vast archangel wings, as if to embrace some holy ark.

And the albatross in Coleridge's *Rime of the Ancient Mariner* was a light-mantled sooty albatross (*Phoebetria palpebrata*), according to Dr. Knowles Kerry, the senior biologist with the Antarctic Division of the Australian Department of Supply. On the eve of his departure in November 1972 on an expedition to the sub-Antarctic Macquarie Island, Dr. Kerry said science had identified the albatross hung round the neck of the Ancient Mariner. The poem was inspired by a real incident when a member of the crew of a ship rounding Cape Horn shot an albatross which ornithologists have now identified as a light-mantled sooty albatross. (This bird is also sometimes called the gray-mantled albatross, which belongs to the same genus as the sooty albatross, *P. fusca.*)

In our discussion of the large and small eggs of living birds we were looking at typical eggs. Even in a single species there's often a considerable range of size in eggs, and occasionally freakish extremes—such as the domestic hen who suddenly delivered herself of an egg weighing eleven ounces, nearly five times larger than the norm. This masterpiece by an ambitious white leghorn is one of the showpieces of the Museum of Curiosities of the Pasteur Institute in Paris. Another hen laid an egg weighing only a quarter of an ounce. Such abnormalities are caused by a temporary disturbance or disease.

Eggs of the same species may also vary from country to country. The eggs of the great white egret (*Egretta alba*) are usually larger in Europe than they are in India. No one knows whether the difference is due to hereditary or environmental factors.

Although, as we've seen, large birds lay large eggs and tiny birds lay tiny eggs, the larger the bird, the smaller the weight of the egg relative to the bird. A small bird, such as a finch, will lay an egg almost one-ninth of her weight, but a large bird, such as an ostrich, will produce an egg weighing only one-sixtieth of her body weight. This relative scale is perhaps best shown among seabirds: The large albatrosses lay eggs that are about 6 percent of their body weight; the medium-size fulmars, eggs that are about 15 percent; and the eggs of the smallest petrels are about 20 percent of their body weight.

Possibly the bird holding the record for laying the relatively largest egg is the kiwi (*Apteryx* spp.); these hen-size birds of New Zealand are widely credited with laying eggs a pound in weight, about one-quarter of their body weight. However, not long ago a female *Apteryx australis mantelli* was killed on the point of laying. The Dominion Museum, Wellington, found that the bird weighed seven pounds and her single elongated white egg one pound. If this proportion (14 percent) is verified by studies of other kiwis, then the kiwi would be somewhere around the scale of the medium-size fulmars (15 percent).

As a general rule precocial species, whose young when hatched are active, lay larger eggs than altricial species, whose young are at first naked and helpless. For instance, the precocial guillemot or common murre (*Uria aalge*) is about the same size as the altricial common raven (*Corvus corax*), but the egg of the guillemot is about twice as large as the raven's.

And birds who lay large clutches of eggs usually lay smaller eggs than birds of similar size who lay smaller clutches. One exception to this rule is the European cuckoo (*Cuculus canorus*), which is much larger than the skylark (*Alauda arvensis*) but lays an egg about the same size. The cuckoo usually lays its eggs in the nests of smaller birds, such as hedge sparrows, robins, pipits and wagtails, and, of course, its eggs must therefore be as small as those of the host bird. Because the cuckoo eggs are the same size and also closely resemble those of the host bird, the host bird hatches them along with its own eggs.

Seemingly fragile, the oval shape of most eggs gives them quite remarkable strength. Embodying the same principle as that of the arched stone bridge, oval eggs will bear extraordinarily heavy weights on their convex surfaces before breaking. Scientists have found that the average weight needed to crush a fowl's egg was about 10 pounds. They had to pile 13 pounds on a turkey's egg, 26 pounds on a swan's egg and 120 pounds on a tough-shelled ostrich egg before they broke. At the other extreme, it took 3½ ounces to break a tiny finch's egg. You can test the strength of a hen's egg yourself. A man cannot break one by cupping it in the palm of one hand and squeezing. Little wonder that someone once acclaimed the egg as the best piece of packaging design in the world!

Birds' eggs come in widely divergent shapes, which oologists* have tried to systematize. The *Handbook of North American Birds* recognizes four principal shapes and three relative lengths of each, taking "spherical" as the short elliptical. These are : long elliptical, elliptical, spherical, long subelliptical, subelliptical, short subelliptical; long oval, oval, short oval; long pyriform (pear-shaped), pyriform, short pyriform. "Ovate" is regarded as a synonym of "oval," and "elongate" of "long."

Most birds lay eggs resembling the shape of the domestic hen's egg—that is, oval-shaped with one end slightly broader than the other. Some owls, kingfishers, penguins and titmice lay round or nearly round eggs. Eggs of grebes, pelicans, cormorants and bitterns are pointed at each end. Many hawks and eagles lay eggs that are almost spherical (short oval). Some species, such as doves and goatsuckers, lay eggs that are approximately ellipsoid—that is, a solid figure corresponding to an ellipse. Swifts, hummingbirds and swallows lay long, elliptical eggs. Curlews, plovers, sandpipers and gulls lay pear- or coneshaped eggs that taper sharply from the broad end.

Some of these variations in shape undoubtedly help the species that lays them. For instance, some shorebirds, such as auks or guillemots, lay pyriform eggs. Because of their shape,

---

* Oology is the study of birds' eggs, and oologists are those who study them.

these eggs, when bumped, don't roll but rotate around the pointed end. Guillemots lay their single egg or pair of eggs on the flat narrow ledges of cliffs, where oval eggs might be knocked over by the birds or blown over by the wind.

Some other shorebirds, such as avocets, lay three or four either pyriform or oval eggs that fit snugly together with the narrow ends pointed inward like the slices of a cake, thus making the least space to be covered by the brooding mother. In an experiment Carl Welty disarranged the four eggs of a killdeer (*Charadrius vociferus*), and each time the bird moved them back so that the pointed ends pointed inward.*

For some shorebirds, too, there is a survival advantage in eggs that resemble the pebbles among which they are laid. This camouflage in shape is usually augmented by camouflage coloring in the form of fleckings and markings. Camouflage in egg color is discussed more fully later in this chapter.

Professor Carl Welty thinks that the shape of a bird's egg is probably determined while it is in the magnum of the oviduct (the second and largest section), and that the shape of the pelvic bones influences the shape of a bird's egg. Thus, birds with deep pelvises are more likely to lay nearly round eggs while birds with compressed pelvises are more likely to lay elongated eggs.

The surface texture of eggs is much like that of photographs—they come with either a dull mat finish or with a shiny or glossy one. The eggs of most birds have a dull mat finish like those of the domestic hen; but woodpeckers and most tinamous lay eggs with a glossy, porcelainlike finish. Most birds' eggs, too, are smooth like those of the domestic hen; but ostriches, storks and toucans lay eggs with deeply pitted surfaces, and emus, cassowaries and some chachalacas lay eggs with rough or corrugated surfaces. Other birds who lay eggs with rough surfaces are grebes, boobies, flamingos and

---

* *A Life of Birds.*

some cuckoos. Some birds, such as grebes, boobies and anis (New World nonparasitic cuckoos), lay eggs with a chalky surface; but the eggs of many waterfowl, particularly ducks and geese, have a greasy or oily surface, which presumably may help to shed water.

As a general rule, large eggs have thicker shells than small eggs. Thickness of shell, too, is influenced by how roughly the bird may treat its eggs. Clumsy African francolins trample all over their eggs, which, fortunately, are so thick-shelled that ornithologists claim you can almost bounce them off a wall. The shell of a francolin egg amounts to over 28 percent of the total weight of the egg. Birds who are gentle and light-footed, such as ducks, sandpipers and doves, usually have thin-shelled eggs.

The number of eggs that particular birds lay in each clutch varies widely. (A clutch is, of course, the number of eggs a bird lays at a particular time. Some birds incubate two or more clutches during the breeding season.) Some birds lay one egg only, and others a dozen or more. Those who lay one egg include the larger penguins, petrels, albatrosses, guillemots, some large vultures, most puffins, the bigger doves and some swifts and goatsuckers. But ducks and titmice may lay from 8 to 12, and the bobwhite quail (*Colinus virginianus*) lays about 12 or even 16. As a general rule, each species lays the number of eggs that on average will produce the greatest number of survivors in a brood. Thus, the common murre or guillemot, which nests on cliff ledges and has few predators, need lay only one egg to ensure the future of its race. And the larger penguins, in spite of adverse conditions in the Antarctic, can perpetuate their species with a single egg because the density of the colonies provides some protection against attacks by predators such as skuas (gulls). But the bobwhite quail, which nests on the ground, runs much greater risks of having its eggs taken by predators or trod upon by animals and seeks safety, as it were, in numbers. Some perching birds lay fewer eggs than many birds that nest on the ground. This would appear to be because eggs laid in nests in trees are usually better protected than those laid on the ground. However, there's no firm rule that the number of eggs depends only on the location of the nest; the number of eggs in a clutch depends on many things, such as enemies, whether there is plenty of food or little, and so forth; clutch size depends also on the number the bird can cover with its body during incubation. For instance, some ground-nesters, such as the northern gannet (*Sula bassana*) and the sooty tern (*Sterna fuscata*), lay only one egg usually; just as in the large penguin colonies, there's protection for the eggs in the immense numbers of brooding birds.

Also, the number of eggs that a hen lays has been determined in the course of evolution by how much food is available. Thus, migrating birds breeding in northern Canada or England and Scotland during the summer have up to 16 hours of daylight in which to gather food for their young, whereas a bird nesting in the tropics will have only 12. Birds breeding in spring and summer in northern Canada or England and Scotland will generally lay more eggs in a clutch than birds of the same size breeding in the tropics or subtropics. A striking illustration is that in Britain the European goldfinch (*Carduelis carduelis*) lays on average 5 eggs in a clutch. This bird was taken to Australia about a hundred years ago; there the European goldfinch lays on average 3.7 eggs in a clutch. The shrinkage in clutch size is probably due to the shorter length of the day in Australia for food gathering. The size of clutches tends to rise when the food supply improves. Thus, during a plague of voles in Scotland a few years ago, short-eared owls (*Asio flammeus*) stepped up their average clutch size (of three to five eggs) to from seven to nine. And when there's a lemming-population explosion in the Arctic, both the snowy owl (*Nyctea scandiaca*) and the rough-legged buzzard (*Buteo lagopus*) increase their clutch sizes. The introduction of the European rabbit to Australia led to an increase in the average clutch size of the little eagle (*Hieraaetus morphnoides*), the wedge-tailed eagle (*Aquila audax*) and the Australian brown goshawk (*Accipiter fasciatus*). But with the introduction of myxomatosis, a virus disease that destroyed most of the rabbits, the clutch sizes declined. Much the same sort of thing happened in southwestern England when myxomatosis broke out; during one year many buzzards did not breed.

Some birds are determinate layers—that is, they will lay a set number of eggs and cannot be induced to lay any more. A determinate layer, such as the barn swallow, will lay its normal clutch of five eggs, and if for the sake of experiment you take one away from the normal clutch of five eggs, the bird will be quite content to hatch four eggs. If the bird has laid three eggs and you add one, the bird will continue to lay two more eggs. Birds that lay fixed clutches include the American crow (*Corvus brachyrhynchos*), doves, shorebirds, and large birds of prey. An ornithologist failed to persuade a herring gull to lay more than its typical clutch of three eggs, although he both removed and added eggs.

As opposed to birds who would appear to be able to "count," other birds—indeterminate layers—can be induced to lay many more eggs than they normally do. A yellow-shafted flicker (*Colaptes auratus*), which usually lays 6 to 8 eggs, was persuaded to lay 71 eggs in 73 days. The experimenter removed each egg as it was laid, but always left one behind. A house sparrow (*Passer domesticus*) usually lays 3 to 7 eggs in a clutch. In one experiment an ornithologist took away each egg as it was laid; the hen laid 51 before she gave up. A wryneck (*Jynx torquilla*), which lays 7 to 10 eggs, was induced to lay 62 eggs in 62 days. In another experiment an ostrich, which normally lays about 20 eggs, was induced to lay 65 eggs over three months. The most famous of all indeterminate layers, of course, is the domestic hen, a descendant of the jungle fowl, which normally lays 30 to 40 eggs. Some domestic fowls have laid up to 352 eggs in 359 days. And one record-breaker laid 361 in 365 days. A domestic duck has done even better, with 363 in a year.

But even indeterminate layers will not be persuaded to go on laying eggs if you leave all the eggs in the nest. Presumably the feel of the eggs against the bird's breast tells it that it has laid enough.

As we've seen, in the course of evolution most birds acquired colored eggs. Some birds, however, lay white or near-white eggs, which was presumably the color of the eggs laid by the first birds to develop from reptiles. (Reptile eggs today are white or near-white.)

Birds that lay white or near-white eggs include many complete orders, such as the albatrosses and petrels; the parrots; the owls; the kingfishers; the motmots; and the woodpeckers. Some smaller groups also lay white eggs. The nesting habits of some birds make their eggs less conspicuous than those of other birds—they lay them in holes, or they begin to sit on the eggs after they've laid the first egg, or they cover the nest with down or vegetation when they leave the nest. Birds that lay their eggs in holes include swifts, owls, petrels, some doves, parrots, woodpeckers and kingfishers. Not only don't they need to camouflage their eggs with color, but for all these there could be an advantage in having brilliant white eggs, because the brooding birds can see them more easily in dark nests and holes.

Birds that build open nests, lay white eggs and begin to incubate soon after they lay the first egg include some doves, herons, hummingbirds, owls and grebes. Birds that build open nests and conceal their white eggs with down or vegetation when they leave the nest include some ducks, geese, grebes and many gallinaceous species (grouse, quail, pheasant, etc.).

Birds' eggs also come in an amazing range of colors and markings. Astonishingly, the colors come from two main pigments, red-brown and blue-green. Blue-green pigments are probably derived from the bile, and red-brown pigments probably from the hemoglobin of the blood.

The markings, as varied as the birds themselves, include scrawls, such as on the eggs of buntings; marbling, as on the eggs of nightjars; and mottling in many species. The larger end, which is laid first, is usually the most heavily marked. Blue-green eggs tend to be uniform in color or tint, such as the eggs of cormorants, thrushes, herons and starlings.

Unusual eggs include the almost black eggs of the chilean tinamou (*Nothoprocta perdicaria*) and some very black eggs that have been found in South America; the bird that lays them has not yet been identified.

By far the greater number of birds lay colored eggs, and for good reasons. One is that in the course of natural selection, eggs which were colored or marked tended to blend with the surroundings and to confer some protection on the species that laid them. Thus, the greenish buff, black-flecked eggs of the ground-nesting lapwing (*Vanellus vanellus*), one of the plovers, blend so perfectly with their surroundings that only the keenest-eyed naturalist can ever find them. Once a man was told to look along a furrow in a ploughed paddock in England in which there were six of this plover's nests. When he reached the end of the furrow he'd walked on one nest and failed to spot the other five! Again, the eggs of ducks echo the green of the reeds in which they nest. And the eggs of the red grouse (*Lagopus scoticus*), which are yellowish-white with blotches of reddish-brown or very dark brown, merge perfectly with the purple heather in Scotland where it nests. It is the same with many other ground-nesters, such as the Canada or spruce grouse (*Canachites canadensis*), whose buff-colored eggs, boldly blotched and spotted with shades of brown, blend perfectly with the ground in the coniferous forests where the birds nest.

Camouflage in eggs takes extraordinary forms with some birds. Burchell's courser (*Rhinoptilus chalcopterus*) of Africa builds no nest and lays its eggs on the ground among the rounded droppings of animals. The eggs, with their thick network of fine black lines or blotches, resemble the droppings closely.

The European nightjar (*Caprimulgus europaeus*), like others of its family, makes no nest and lays its eggs directly on the ground. Laid usually near dead wood, the eggs, with their black splotches and lines, merge into the background.

There's protection, then, for most eggs if they're inconspicuous—paradoxically, protection for the eggs of some birds probably comes from their bold, conspicuous colors and unpleasant flavors. Predators learn to associate the bold colors of the eggs with their unpleasant, bitter taste.

Dr. Hugh B. Cott and colleagues at the University Museum of Zoology at Cambridge, England, checked on the taste of the eggs of 134 species of birds and found that many of the brightly colored eggs had an unpleasant bitter taste—at least, unpleasant by our standards. The eggs with the bitterest taste were also the most striking in whiteness or color, usually bright white or white spotted with red and green. It has since been observed that egg-eating animals, such as snakes, hedgehogs and mongooses, usually leave brightly colored eggs alone. Thus, coloration could be another form of protection for birds' eggs; a predator would only have to spit out a few repulsive-tasting eggs before learning to associate the warning color with the unpleasant taste. This is, of course, speculation. We do not know if mammals see color as we do. Many lizards have true color vision; some snakes may have it, too.

Birds that lay unpleasant-tasting eggs include tree sparrows (*Passer montanus*), house wrens (*Troglodytes aedon*), litte terns, linnets, tits and swallows. Among the bitterest were the eggs of the common puffin (*Fratercula arctica*). But Dr. Cott and his colleagues found the eggs of house wrens almost as nauseous—ironically, because their eggs and those of other North American wrens, with blotchings of lilac or brown on a whitish ground, are among the most beautiful of all in appearance.

But if striking and distinctive eggs had the worst taste, Dr. Cott found, at the other extreme, that many of the best-tasting eggs are those with drab colorings and markings.

Egg color is almost certainly inherited. The famous British ornithologist David Lack points out in his book *The Life of the Robin* that robins' eggs of an unusual type, with a faint bluish tinge and minute red spots, were found in a Kent orchard in the years 1909, 1912 and 1922. The average life of a robin is about a year, so it is most unlikely that the same bird laid the three clutches; the 1922 clutch was certainly laid by a descendant of the bird who laid the first clutch, and it's highly probable the 1912 was also.

The eggs laid by most species of birds are usually so uniform in coloration that an expert can readily identify them. In England not so long ago an oologist identified over 4000 out of 5000 eggs in a few hours to give evidence of illegal egg collecting. However, considerable variation can occur in some birds. For instance, the eggs of the South American tinamous have been known to vary from pale primrose to sage green or light indigo, or from chocolate brown to pinkish-orange. Of all birds, probably the guillemots show the widest variation in the eggs they lay. Their large pyriform eggs, according to one ornithologist, may have a ground color of deep blue-green, or bright red, or brownish-yellow, or pale blue, or cream, or white! This ground color may be marked with blotches, spots, zones, or patterns of hatched lines which also vary in color from light yellowish-brown to light red, rich brown or black! Or the egg may have no markings at all. This extraordinary variety is probably not accidental; guillemots nest in great colonies, and distinctive eggs could help the bird to find its own.

All eggs in a particular clutch will usually resemble each other closely, but a hen's second clutch will generally show some variation in markings from the first.

Do birds recognize their own eggs? Well, as we have seen, the guillemot almost certainly does. (The guillemot not only *recognizes* its own eggs but in one experiment scientists placed the eggs at various distances from the nest and the bird rolled the eggs back to its nest from distances up to 16 feet.) Many European robins also know their own eggs; David Lack found that many robins deserted a nest when a cuckoo's egg was laid in it. But great black-backed gulls (*Larus marinus*) and herring gulls (*Larus argentatus*) are more easily fooled; they will accept any more-or-less rounded object about the same size as an egg. Zoologists have successfully persuaded the unfortunate birds to sit on rubber balls, watches, cans and even small bricks with roughly rounded corners. And gannets have readily accepted Coca-Cola bottles!

Experiments with other birds have shown that many species will sit on almost anything you place in their nests, such as light bulbs, golf balls, mollusk shells, brightly painted blocks and so on. Geese have been induced to try to hatch ostrich eggs. A European oyster catcher (*Haematopus ostralegus*) tried to incubate a gigantic china egg almost as large as itself! You must not think that European robins are necessarily brighter than gulls or other birds. Obviously, in the course of evolution there has been survival value in European robins recognizing eggs laid by parasitic cuckoos. But gulls have obviously never had to contend with cuckoos, and if a pebble or other strange object did get in among a gull's clutch of eggs, it wouldn't matter a great deal.

The amount of yolk in birds' eggs—the main source of food (do you remember Wilhelm Bölsche's "thickly buttered bread"?)—varies greatly among species. It can range from 12 percent by weight in the golden eagle to about 60 percent by weight in the megapodes, the mound builders of the Australian-Asian region. But those are extremes; as a general rule those birds whose eggs have a relatively low percentage of yolk—from about 15 percent to about 20 percent—are all birds who produce altricial young—that is, young who are blind, naked and helpless when first hatched. Typical altricial birds are the passerine or perching birds, including doves, pigeons, starlings and most songbirds.

Birds whose eggs have a high percentage of yolk by weight are the precocial species whose young are hatched with eyes open and with down and are able to leave the nest in a day or so. The percentage of yolk can range from over 32 percent to 50 percent in most gallinaceous species, which include the megapodes, grouse, pheasants, turkeys, guinea fowls, rails and domestic hens. Other precocial species include the ostrich, and the emu.

Young megapodes are independent of their parents from the start. Ducklings follow their parents and find their own food. So do newly hatched plovers. The young of other precocial species follow their parents and are shown food, or they follow their parents and are fed by them.

Alexis L. Romanoff and Anastasia J. Romanoff, in their classic work *The Avian Egg*, compiled a table of the percentages of yolk in various precocial species and observed that large eggs had proportionately less yolk than small eggs. An ostrich egg, they found had 32.5 percent yolk, and a plover's egg—at the bottom of the table—had 52.9 percent. The reason, they deduced, was that the smaller the egg, the greater, relatively, was its surface. "A large surface area and a thin shell tend to permit rapid loss of heat," they write. "As if to compensate, the embryo is provided with larger amounts of food substances of high energy value in the form of yolk."

Incidentally, the Romanoffs give the percentage of yolk in the average domestic hen's egg as 31.9 percent. "By weight, the hen's egg is roughly six parts albumen, three parts yolk and one part shell." They add that hens' eggs vary quite widely—even the eggs laid by a particular hen.

Variation—even oddity—sometimes extends to shape in hens' eggs and, too, in the eggs of wild birds.

Hens' eggs shaped like cucumbers and sausages have been recorded. Double eggs with linked shells occur less frequently. Much more familiar are double-yolk eggs. A record double-yolker weighed just over seven ounces. Investigating this phenomenon, which occurs chiefly among young pullets, a scientist computed that a double-yolk egg appears once in about every 530 eggs. The odds against triple-yolkers are very much greater—one egg in every 5000.

35

INCUBATION

BIRDS' EGGS must be kept at a comparatively high temperature if they are to hatch. The average incubating temperature is about 34° C. (93° F.). An ornithologist inserted thermocouples into incubating eggs of 30 different species of birds, representing 11 orders, and found the average temperature to be this figure of 34° C.

In order to apply more heat to their eggs, most incubating birds develop "brood patches" to help the transfer of heat to the clutch. These are areas of bare skin on the bird's belly that look red and inflamed to our eyes because the skin has become spongy and richly supplied with blood vessels. (Feathers are, of course, a splendid form of insulation; heat moves slowly through still air, and feathers trap lots of air. An ornithologist found that a house wren (*Troglodytes aedon*) was able to transfer 10° F. more heat to its eggs after it had shed its breast feathers.) Some birds have a single brood patch. These include many perching birds, birds of prey, grebes and pigeons. Some seabirds, such as auks and skuas, have two brood patches. Other birds, such as waders, gulls, gallinaceous birds (including the domestic fowl), have three. Some have none at all;

they include ducks, geese, cormorants, gannets and penguins. Gannets and penguins make up for the omission by highly developed blood vessels in their webbed feet for the warming of their eggs. Gannets warm their eggs by standing on them. Murres and penguins hatch their eggs by holding them on top of their feet. Ducks and geese make their own brood patches, as it were, by plucking the down feathers from their breasts and using them to line their nests. By contrast, some sea birds that lay their eggs on tropical coral cays, such as sooty terns (*Sterna fuscata*), spread their wings to protect their eggs from the fierce sun.

With some birds incubation starts with the laying of the first egg; these include hornbills, loons, grebes, pelicans, herons, storks, crows, eagles, hawks, cranes, many gulls, parrots, owls, swifts, hummingbirds and some perching birds. Birds that don't start incubating until the last egg has been laid include ducks, geese, gallinaceous birds and most perching birds. In other birds incubation starts before the laying of the clutch is completed; these include the ostrich, rheas, rails and woodpeckers.

What distinguishes birds from their reptilian ancestors is the degree of cooperation between parent birds. Many paired birds share the building of nests; most share the incubation. Two American ornithologists, Josselyn Van Tyne and Andrew J. Berger, in *Fundamentals of Ornithology*, tell about a survey of some 160 families, representing the majority of living birds, and found that in 54 percent of the families both sexes usually incubate the eggs. They found that the female alone incubates the eggs in about 25 percent, the male alone in about 6 percent, and in about 15 percent it was either the female or the male alone, or both birds—that is, in this 15 percent there was no regular pattern. It has been suggested that incubation by both sexes was the primitive method among birds.

Among the more remarkable males who incubate the eggs are the kiwis (three species), the emus, rheas, phalaropes and emperor penguins.

No bird in the world incubates its egg under harsher conditions than does the emperor penguin. Weighing between 50 and 100 pounds and standing 3 feet high, the emperor penguin is the world's largest and heaviest seabird. It has a slender downward-curving bill and orange-yellow patches at the sides of its neck. The emperor penguins are probably the only birds who never set foot on land; they spend their time either at sea or on the sea ice below Antarctic ice cliffs. Two important factors impose harsh conditions on the male, who incubates the egg: He must choose a time during the winter when the sea ice is frozen hard; and the eggs must hatch in July and August when the sea ice is melting, so that the young can find food close at hand. For these reasons the breeding cycle begins in midwinter, when the female lays her single egg. The male immediately moves it onto the top of his feet, where it will be kept warm by a drooping fold of his belly skin, aided by the specially adapted warm feet described earlier. Males huddle together to conserve heat, each holding his single egg on his feet for eight or nine weeks. Throughout the whole time he dares not expose the egg to the cold; letting it roll onto the ice for even a few seconds would almost certainly kill the embryo. Throughout the dark Antarctic night the birds, sometimes as many as 6000 in a single colony, crowd together; temperatures may go as low as $-60°$ C. ($-77°$ F.), and the mean wind velocity may be from 10 to 22 miles per hour, with an occasional blizzard. Throughout the eight or nine weeks the male cannot eat but must live off the blubber that makes up a third of his body weight.

Meanwhile, the females have gone to sea to feed. They may have to cover as much as 50 or 100 miles of ice before they reach the sea.

There's a graphic account by a member of Robert Falcon Scott's Antarctic expedition of 1911. Three members of the expedition walked from Cape Evans to Cape Crozier, 67 miles away, to obtain the eggs of the emperor penguin. Here is what one of them, Apsley Cherry-Garrard, wrote of the unique sight, which no other man had seen before:

> We saw the Emperors standing all together huddled under the Barrier cliff some hundreds of yards away. The little light was going fast: we were much more excited about the approach of complete darkness and the look of wind in the south than we were about our triumph. After indescribable effort and hardship we were witnessing a marvel of the natural world, and we were the first and only men who had ever done so; we had within our grasp material which might prove of the utmost importance to science; we were turning theories into facts with every observation we made—and we had but a moment to give.
>
> The disturbed Emperors made a tremendous row, trumpeting with their curious metallic voices. There was no doubt they had eggs, for they tried to shuffle along the ground without losing them off their feet. But when they were hustled, a good many eggs were dropped and left lying on the ice, and some of these were quickly picked up by eggless Emperors who had probably been waiting a long time for the opportunity.

The chicks hatched in July and August are fed by the males with a secretion from their crops. Then the females return to take care of the young while the emaciated males, which, according to recent research, have lost half their body weight during their ordeal, make their way across the ice to the sea to feed on fish, krill and squid. Two or three weeks later, his strength regained and with about seven pounds of fish in his crop, the male returns to his chick. Thereafter both parents take turns bringing food to the chick, and its growth is rapid.

Fatherhood is also strenuous for the male kiwis of New Zealand. Incubation may take 75 to 77 days. The male leaves the nest only at dusk to feed hurriedly. By the time the incubation is over, he has lost half his body weight.

Male rheas have harems of up to six hens, each of whom lays in the single nest the male has scooped out. The male may have to incubate from 20 to 50 large eggs for 35 to 40 days.

The male emu of Australia is also polygamous and incubates up to 20 eggs laid by his harem for 58 to 61 days.

One of the oddities is the simultaneous incubation in different nests by the European red-legged partridge (*Alectoris rufa*). The hen builds two nests, then lays a clutch in the first nest and another in the second. She incubates the clutch in the first nest, and the male incubates the clutch in the second. Each parent cares for its own young.

In Africa both the male and female common waxbills (*Estrilda astrild*) sit side by side in their dome-shaped nest to incubate their eggs.

Communal incubation is the practice of some species, such as the anis (*Crotophaga* spp.), which build communal nests. In a shallow cup of sticks lined with leaves, several females lay up to 24 eggs and incubate them companionably together. Another communal breeder is the babbling thrush (*Yuhina brunniceps*) of Formosa. Six birds were observed incubating eight eggs.

Among birds the brooding time varies considerably from about 11 days to almost 12 weeks. Perching birds, such as the brown-headed cowbird (*Molothrus ater*), take about 11 days. At the other end of the scale, the royal albatross incubates its eggs from $77\frac{1}{2}$ days to $80\frac{1}{4}$ days.

The instinct to incubate eggs is powerful in many species of birds. For instance, a male bobwhite quail sat on a nest of 13 eggs that failed to hatch for a period of 81 days. And as we saw earlier, many birds will "incubate" almost anything roughly resembling an egg in shape placed in their nests. Lacking eggs, penguins have been observed to try to incubate stones or even lumps of ice. Geese have been persuaded to attempt to incubate ostrich eggs.

Sitting on the eggs is the norm, but, as in all things in nature, there are special cases, which we'll look at next.

36

MOUND BUILDERS

A FEW REMARKABLE BIRDS do not incubate their eggs at all but leave this to the sun, much as their reptilian ancestors did. One is the double-banded courser (*Rhinoptilus africanus*), who lays a single egg on bare ground during the dry season in northern Tanganyika. The parent birds do not incubate the egg but instead take turns standing over the egg to protect it during the daylight hours from the heat of the sun. A closely related bird, the black-backed courser (*Pluvianus aegyptius*—or the Egyptian plover, as it is sometimes called), lays its eggs in sand with only the rounded tops showing. It is reported that the parent birds sometimes wet the sand by regurgitating water over it. Further, the parents use sand not only to protect and incubate the eggs but sometimes to hide the chicks. When danger threatens, the chicks flatten themselves in the sand and the parents kick the sand over them with their feet, burying them temporarily until the danger is over.

Even more remarkable is the Megapodidae family of

birds, known as "mound builders," which are restricted almost entirely to the Australasian region. The name "megapode" means "large foot" and is apt enough; these ground dwellers have powerful feet. The megapodes differ widely in appearance, but they do have one thing in common—they do not build nests and hatch their eggs as most other birds do; instead, they bury them in the ground and incubate them by various means. On Savo, in the Solomon Islands, the jungle fowl (*Megapodius freycinet*) nest in sandy areas where percolating volcanic steam hatches the eggs. Other megapodes in tropical areas bury their eggs in the ground or in heaps of fermenting vegetable material they have scratched together.

Mound builders first caught European attention in 1521 when Antonio Pigafetti, who had accompanied Magellan on his circumnavigation of the world in 1519–1521, wrote an account of a bird he had seen in the Philippines, which he said was about the size of a fowl, laid its eggs in holes in the sand and left them to be hatched by the sun's heat. Later in the same century other accounts were written for wondering readers by other observers but were generally dismissed as tall stories. Some sixteenth- and seventeenth-century naturalists flatly refused to list the megapodes at all!

Today we know that there are seven genera and about twenty species of megapodes. Two genera are confined to the Australian mainland, and the others live in the islands, to the north and west—in the Nicobars, on islands off Borneo, in the Philippines, Marianas, Celebes, New Guinea, Solomons, the New Hebrides and central Polynesia. For convenience the megapodes can be sorted into three ecological groups. The first and most widespread is that of the junglefowl with three genera. One species, *Megapodius freycinet*, is found with many subspecies throughout the whole range of the birds, including northern Australia. Two genera with a single species each, the maleo (*Megacephalon maleo*) and the painted megapode (*Eulipoa wallacei*) are confined to the Celebes and Moluccas respectively. Junglefowl are small birds somewhat smaller than domestic fowls; they are dark in color and long-legged.

The second group of the three genera of brush-turkeys is more restricted—one, *Alectura lathami*, is confined to the eastern coast of Australia and the other two, *Aepypodius* spp. and *Telegalla* spp., to New Guinea only. The brush-turkeys are larger than junglefowl and use their wings less.

The junglefowl and the brush-turkeys live in tropical or sub-tropical regions of high rainfall.

The third group with one monotypic genus is, perhaps, the most fascinating of these most extraordinary birds. The mallee fowl (*Leipoa ocellata*) is a brown, white and black bird and rather resembles the brush-turkey in appearance. Its habitat is the arid scrublands of the harsh Australian inland. "Mallee" is an Aboriginal term and was used by some Victorian Aborigines for the small, hardy *Eucalyptus dumosa*. The term is now applied to other hardy, low-growing trees and shrubs of the eucalypt family.

Most of our knowledge of these extraordinary birds is due to the seven years' research work of Dr. H. J. Frith, Officer-in-Charge of the Wildlife Survey Section of the Commonwealth Scientific and Industrial Research Organization.

Unthinking writers have sometimes dismissed the mallee fowls—and the other megapodes—as "lazy birds who couldn't be bothered to hatch their own eggs." On the contrary, as Dr. Frith has learnt, the mallee fowls work extremely hard for most of the year in order to reproduce themselves. As he points out in his book, *The Mallee Fowl* (Sydney and London: Angus and Robertson), the bird has an even more difficult task than its cousins in the tropics. In the tropics there is little variation in temperature from month to month and the daily variation is also quite small. But in semi-arid inland Australia, the daily range is great; over the year it will fluctuate from below freezing point at night to well over 100 degrees during the day. Even in summer the daily temperature may seesaw 30 degrees.

In order to use the heat from fermenting vegetable matter, the male bird is strenuously engaged for about 11 months of the year. He spends from two to three hours digging out and replacing the mound every day. And though of a different nature, the strain on the female is just as severe. Throughout the year she lays from 15 to 24 eggs, each about one-tenth of her own weight. She may even lay as many as 35 pale pink eggs. This places a tremendous strain on her constitution, so most mound tending is necessarily left to the male; she only joins him in this work after egg laying is finished.

As we have already seen, the egg weight of birds decreases relative to body-weight increases. If we adopt the scale of a normal bird, then the mallee fowl, which weighs about 63 ounces, should lay eggs of about 2.8 ounces. The eggs, in fact, weigh about 6.3 ounces. The yolk represents about 60 percent by weight. These factors probably account for the very advanced state of development of mallee fowl when hatched. The mallee-fowl chick can run very swiftly within a few hours of hatching.

That a bird builds an incubator is wonderful enough, but that the bird has a built-in "thermometer" is even more remarkable. The male maintains the heat in the 15-to-18-foot-diameter mound at a temperature of 92° F. When, because of fermentation, the heat becomes too great for the eggs, he digs deeply and exposes the chamber to the cool morning air. (See Figure 10.)

Dr. Frith and his coworkers buried instruments in mounds to measure the temperature and record it continuously. The male bird came to the mound each morning,

FIG. 10. *Mallee-fowl cock controls temperature of incubator mound by adding or removing mound material.*

coming a little later each morning as the summer waned. After his digging, the temperature of the mound would drop to 92°. During the rest of the day and night it rose slowly, to be cooled by the bird's excavations the next morning. On some occasions the male would not even allow the female to disturb the nest by laying an egg—presumably because this would disturb the temperature. One bird dug furiously during a sudden thunderstorm in an attempt to correct a depression that would catch too much water and presumably affect the incubation of the eggs. His labors were in vain; the eggs did not hatch. Dr. Frith says that no one, least of all himself, had ever dreamed that the bird could control the temperature of a heap of sand and leaves so accurately. How does he "measure" the temperature?

Probably by probing the mound with his bill, which is very sensitive.

The eggs, which are pale pink when newly laid but are later stained brown by the red soil and the organic matter in the mound, take 50 days to hatch. As astonishing as the male's remarkable abilities is the emergence of the fledgling chick, three feet under the sand, and its struggle up to daylight and life. It may take a newly hatched bird from 2 to 15 hours to emerge. You might expect it would suffocate, but it apparently draws enough air from the pores between the grains of sand. Here again, the innate behavior of the male plays its part; when the eggs are hatching he scratches the sand and loosens it.

When it emerges from the sand, much like a butterfly from a chrysalis, the little bird struggles feebly away from the mound and into the shelter of the mallee. There it gathers strength swiftly. Within an hour it can run well. And within 24 hours it can fly strongly.

There are many marvels in nature, but none more remarkable than birds who build incubators and regulate temperature.

37

CUCKOOS AND OTHER PARASITES

ONE REMARKABLE FORM of incubation is nest parasitism—that is, a bird lays an egg in another's nest for the host to hatch. No doubt, you're already thinking of the European cuckoo (*Cuculus canorus*), which is the most famous of nest parasites. But it is not the only "offender"; about 80 species of birds belonging to several families are completely nest parasites, and some other species are occasionally parasitic—they usually build their own nests, incubate their eggs and care for their young but sometimes lay an egg or eggs in another bird's nest. Many species of ducks do this, and one species, the black-headed duck (*Heteronetta atricapilla*) of South America, is completely parasitic; it lays its eggs in the nests of rails, coots, ibises, limpkins and gulls. Birds that are sometimes parasitic are of great interest to ornithologists because they may offer some clues as to how this extraordinary habit could have originated.

"Cuckoo" means one thing to most people—the miscalled European cuckoo, which is just as much an Asian cuckoo because it is widespread on both continents. But "cuckoo" to ornithologists means a very large family (Cuculidae) of related birds, which is worldwide in distribution with about 127 species grouped in 38 genera and 6 subfamilies. Not all are parasitic, but one large subfamily, the Cuculinae, of 47 genera, is wholly parasitic. It's an Old World family extending from northern Europe through Asia to Australia, New Zealand and the islands of the Pacific.

No cuckoo, however, has been more studied than the European cuckoo, who is the master parasite of them all. She lays her eggs in the nests of over 300 species of birds. Of course, this does not mean that a single European cuckoo will lay her single egg in all 300, or that all females of that strain will. The parasitic egg must, in most instances, closely resemble the host eggs if the deceit is to succeed. What usually happens is that one strain of European cuckoo lays its eggs in the nest of one host species for generation upon generation. Thus we have European cuckoos who lay in the nests of either reed warblers, meadow pipits or wagtails. And the eggs in most instances closely resemble those of the particular host.

The female European cuckoo usually lays her egg in the host's nest in the afternoon, when the host is most likely to be absent, and she lays it in smart time—under eight seconds. Before leaving, she removes one of the host's eggs.

She lays another egg two days later in the same nest. She may lay 12 or more in various nests of the one host species in the breeding season.

That the one female European cuckoo should always choose the one host species is extraordinary. A possible explanation is that when she was reared by her foster parents she was imprinted with the type of egg and nest—and possibly, host—and retained this impression for life.

European cuckoos usually lay their eggs in the nests of passerine (perching) birds and choose half-completed and freshly completed clutches. The host birds—like other passerines—don't start incubation until the last egg has been laid, an important factor in this singular story, as you shall see.

European cuckoos are usually much larger than their hosts—one of the most grotesque instances is of a tiny host perched on the back of a large and clamorous foster child and feeding it. Because they are much larger and, therefore, require more food, the baby European cuckoo would probably starve to death if the hosts were to attempt to feed it and their young. This does not happen, because the baby cuckoo hatches in $12\frac{1}{2}$ days, ahead of the host's eggs, which take 13 or 14 days. Then, moved by instinct, the blind and naked cuckoo clears the nest of everything but itself. On its back is a sensitive shallow depression. If anything solid, such as an egg or young bird, touches this, the cuckoo moves it to the rim of the nest and thrusts it overboard. This instinctive reflex will respond equally to inanimate objects such as acorns.

Having rid itself of competitors for food, the cuckoo prospers—sometimes while the young it has ejected starve to death just outside the nest, ignored by their parents. Nowhere is the inherited behavior of birds more apparent. The host birds are responding to signals or stimuli to feed baby birds *within* the nest. Even if the cuckoo hasn't ejected all its rivals, it would still get most of the food brought by its hosts, for the reason that, in the course of evolution, the young cuckoo has acquired a very large gape, and also a brightly colored mouth and throat, which is more attractive to the feeding foster parents. Konrad Lorenz has pointed out that in a mixed aviary a baby cuckoo becomes a source of "sin" because all the parents neglect their own young most of the time in order to feed it.

New World cuckoos are not usually parasitic; only two species occasionally lay eggs in the nests of other birds. The most famous New World parasitic birds are the cowbirds. The brown-headed cowbird (*Molothrus ater*) lays its eggs in the nests of 160 species of birds and is probably the most successful of New World parasites.

Like some of the parasitic ducks, the cowbirds are of great interest to ornithologists because they offer some suggestions about how parasitism evolved. Cowbirds include species at various stages of parasitism. One species sometimes builds its own nest but more usually takes another bird's (of a different species). Another exhibits nest-building behavior but rarely builds and uses its own nest; more frequently it lays its eggs in the nests of other birds—and then not very successfully; many of its eggs are found on the ground, and often those laid in the host's nest are too numerous to stand any chance of being incubated. Whereas the European cuckoo, which is the most efficient of the parasites, lays one egg only, from 15 to 37 cowbird eggs have been found in a single nest of certain ovenbirds.

No cowbird ejects its host's young from the nest. Nor do the honey guides of Africa, with the exception of one species, *Indicator indicator*. But the blind nestlings of at least two species of honey guides have an equally lethal habit; when hatched they are equipped with two needle-sharp hooks on the tips of their beaks with which they bite and tear their foster brothers until they die and are removed, presumably by the foster parents. The "murderous" youngsters lose their hooks when they are about two weeks old.

Honey guides are better known for their practice of guiding with excited cries men, honey badgers and some other mammals to the nests of bees. When men or mammals break open the nest, the honey guides share in the feast. Honey guides were studied on 23 guiding trips; the longest was for 750 yards and lasted for 28 minutes; several were from 250 to 350 yards and lasted from 9 to 16 minutes.

From our human viewpoint, parasitic birds appear cruel

and immoral—so much so that some people remove cuckoo and cowbird eggs from nests. But they would only be cruel and immoral if birds were like ourselves and did these things consciously.

This account has, perhaps, given the inaccurate impression that every parasitic bird is 100 percent successful. But some hosts are much more sensitive than others; they either desert their nest if a cuckoo lays in it, or eject the cuckoo's egg. Moreover, one individual parasitic bird may be permanently attached to the wrong host, who rejects all or most of the parasite's eggs. Or a parasite may even be too successful at first and in time leave itself with too few hosts to parasitize. For instance, a single European cuckoo was observed year by year in its parasitizing of a population of reed warblers (*Acrocephalus scirpaceus*). In the first year the cuckoo laid its eggs in 14 nests, or a percentage of 29 percent. Note now the progressive increase in parasitism and *decrease* in the number of nests in succeeding years: 15 nests (40 percent); 12 nests (50 percent); 11 nests (73 percent); 9 nests (67 percent); and 8 nests (88 percent).

Parasitism should be thought of as a form of incubation that some birds—nearly 80 species in all—have adopted with varying degrees of success.

38

MAMMALS THAT LAY EGGS

WE HAVE ALMOST come to the end of our story. Our last chapter is about two of the most unusual of the egg-layers. They are primitive mammals and as such are of tremendous interest in the evolutionary story. Somewhere in the Age of Reptiles the first primitive mammals evolved, and some, like their Australian descendants of today, the platypus (*Ornithorhynchus anatinus*) and the spiny anteater (*Tachyglossus aculeatus*), laid eggs and suckled their young. (The scientific name of the spiny anteater means "swift, prickly tongue," and that of the platypus means "birdlike snout like a duck.")

On August 29, 1884, a young British zoologist at a Queensland sheep ranch excitedly drafted a cable to be transmitted to a meeting of the British Association for the Advancement of Science, which was meeting in Montreal, Canada. W. H. Caldwell, the 24-year-old Cambridge zoologist, wrote four momentous words: MONOTREMES OVIPAROUS OVUM MEROBLASTIC. Translated into the language of the layman, the message read: "The platypus and spiny anteater (echidna) lay eggs. The egg has such an amount of yolk that the embryo first develops as a plate sitting on top of the yolk (just as in a bird or reptile egg)."

When read in Montreal, the cable caused a ripple of excitement because it resolved what had been a hot controversy for over 60 years. Caldwell had gone to Australia to solve it and did so in the Burdekin River district of Queensland. He found the eggs in the pouch of a spiny anteater and an egg about to be laid by a platypus.

About the same time as Caldwell found his eggs, Professor Wilhelm Haacke of Germany also obtained final proof of the egg-laying habits of the platypus and spiny anteater—Caldwell did so during the second week of August 1884, and Haacke on August 25 the same year. Haacke was examining the pouch of a female spiny anteater in South Australia and found an egg in it. His surprise was so great that he crushed it between his fingers!

The first settlers at Botany Bay, Australia, saw the platypus—the Greek-derived name means "flatfoot"—on the banks of the Hawkesbury River in New South Wales. Lieutenant Colonel David Collins wrote in his journal in 1798 what is probably the first written account of this remarkable fur-bearing creature with a beaverlike tail and webbed feet. In his account, Collins said that "the most extraordinary circumstances observed in its structure was its having, instead of the mouth of an animal, the upper and lower mandibles of a duck. By these it was enabled to supply itself with food, like that bird, in muddy places . . . while on shore its long and sharp claws were employed in burrowing; nature thus providing for it in its double or amphibious character." *

Superficially Lieutenant Colonel Collins was right. The snout resembles that of a bird, but instead of being hard and horny it is soft and leathery. It combines the nose and lips of other mammals and is charged with sensitive nerve endings on which the platypus relies entirely in its underwater search for

---

* Collins' *New South Wales*, published in 1802.

FIG. 11.   *When swimming underwater the platypus relies on smell to find its prey. Its eyes and ears are shut within facial furrows.*

food, because when it submerges, both eyes and ears are shut within facial furrows. At the base of the snout there is an extra flap of a rubbery texture which acts as an eyeguard when the animal is underwater. Once described rather aptly as looking "like a hot-water bottle," the 20-inch-long platypus uses only its front legs for swimming, paddling with one after the other. (See Figure 11.)

(The hypersensitive snout has caused the death of at least one platypus. During the war Sir Winston Churchill expressed the wish for a platypus, and one was dispatched to him. It survived the dangerous voyage until the ship was only a few days out from England, when depth charges were dropped against a submarine. The platypus died of shock—its sensitive snout was not evolved to cope with depth charges.)

Other reports soon followed from Australia, to be received skeptically in Europe. The first actual specimens sent to Europe about this time also evoked skepticism. Some scientists assumed they were fakes and tried to pull the bills, which they thought were sewn on, from the bodies. (Clever fakes had been foisted on travelers and sailors in the East, such as the "mermaid," a mummified monkey's head fixed onto the skin of a large fish.) Careful examination of the specimens proved, however, that this strange creature did in fact exist. Further examination revealed another feature as extraordinary as its bill: Like birds and reptiles, the creature had only one vent for purposes of excretion and reproduction—hence the name "monotreme." Reports from the Australian colonists that the creature laid eggs were stubbornly disbelieved by many scientists of the day, who maintained that all mammals were born alive and that because this creature was obviously a mammal, it could not lay eggs.

The discovery in 1824 of mammary glands in the platypus—admittedly rather rudimentary glands, being large pores from which milk oozed while the platypus fed its young lying on its back—did not help to resolve the controversy.

The dispute about the platypus wasn't helped by the discovery of its relative, the spiny anteater, which was earlier called the "echidna" and also "native porcupine." It had many similarities with the platypus, including a single vent.

The discovery of the first spiny anteater was recorded in the log of the *Providence*, which was under the command of Captain William Bligh. The entry for February 7, 1792, gives a good account of the Tasmanian spiny anteater (*Tachyglossus setosus*), which is smaller than the mainland one:

> Lieutenant Guthrie in an excursion today killed an animal of very odd form. It was 17 inches long, and the same size around the shoulders to which a flat head is connected so close

that it can scarcely be said to have a neck. It has no mouth like any other animal but a kind of duck bill, two inches long, which opens at the extremity where it will not admit the size of a small pistol ball. It has no tail but a rump not unlike a penguin, on which there are some quills about an inch long, as strong as those on a porcupine. These quills, or prickles, are all over its back amidst a thick coat of rusty brown hair; but the belly is of a light greying colour.

What Lieutenant Guthrie could have added is that the animal has a long, sticky tongue, which it uses to lick up ants, termites and other insects.

Before long the scientific world was split into three main schools. One argued that the platypus was a mammal and that it must be viviparous—that is, bring forth living young. The second school was convinced that it was egg-laying, or oviparous. The third asserted that it must produce eggs, which, however, hatched within the parent, thus making the animal ovoviviparous.

We know today, of course, that the platypus and the spiny anteater are monotremes, primitive mammals with an ancient history reaching back, perhaps, to the beginnings of mammals in the Jurassic period some 180 million years ago. About that time some reptiles were developing mammalian characteristics, and several different lines of such mammallike creatures had evolved. (About 120 million years ago one of these lines developed branches; one gave rise to marsupials and placental mammals, including ourselves.) It is not true that the Australian monotremes are "missing links" in the line of evolution. They are also much changed from their ancestral types. What they are, however, are "living fossils." Australia, as we have seen, is a great natural reserve for archaic forms, where survivors of early forms of life have been deep-frozen, so to speak.

The star exhibits in the colossal living natural-history

museum that is in Australia are the two monotremes and the marsupials, primitive pouched mammals whose young are born in an immature state. According to the fossil evidence, small marsupials probably lived in the Americas, Europe and Australia but went under eventually before the challenge of the more efficient placental mammals—except in Australia and in small pockets of opossums in North and South America. In the struggle for existence, these more highly developed mammals finally beat the more primitive egg-laying and pouched ones, although initially the two were reasonably well matched.

There were a number of reasons for the marsupial decline—probably the most important was the superior intelligence of the placental mammals. So the monotremes (egg-layers) and nearly all the marsupials died out in the rest of the world. In Australia the marsupials were luckier. When, about 60 million years ago, the land bridges between Asia and Australia were broken, the large placental mammals could not follow there. In their Australian sanctuary, without competition from better-equipped animals, and particularly from large predators, monotremes such as the spiny anteater and the platypus persisted, and the marsupials branched out into many forms not found elsewhere. In this vast sanctuary, the marsupials went on developing to fill all ecological niches and parallel the development of the placental mammals. Kangaroos evolved to eat grass and take the place filled elsewhere by placental grass-eaters, such as sheep, cattle, oxen and deer. Koalas and possums took the places occupied elsewhere by sloths and monkeys; gliding possums took that of flying squirrels; wombats that of marmots; marsupial wolves (*Thylacinus cynocephalus*) and marsupial cats that of true wolves and cats; the Tasmanian devil (*Sarcophilus harrisii*) that of hyenas; and marsupial rats and mice that of small carnivores or insectivores, such as shrews.

It is disappointing that the fossil record of marsupials is scanty. They are thought to have developed in North or South America and spread to Europe and Australia. Their ancestors are believed to have been the docodonts, which were one of a number of primitive mammals in the Jurassic period. They were presumably small insect-eaters.

What is so fascinating about the monotremes is that they show many reptilian features. Like reptiles they have single vents and they also lay small, leathery eggs with parchment-like shells, which resemble closely those of reptiles. Like reptiles, they have ribs in the neck, lack an important cross section between the cerebral hemispheres and cannot maintain a constant body temperature. Yet, incontestably, they are mammals, because they have hair and they suckle their young. They lack teats but can exude milk from pores that are highly modified sweat glands.

An important difference between the spiny anteater and the platypus is that the female spiny anteater has a pouch in which she carries the eggs until they hatch.

Growing as long as two feet, the spiny anteater must be one of the world's strongest animals for its size. It is also, as we shall see, the Houdini of the animal world. It performs extraordinary feats of strength, tearing rotten logs apart and pushing rocks aside in its search for food. Mr. Harry Frauca, the Australian naturalist, wrote in a magazine article that he has had to use a pick or geologist's hammer to break up concrete-hard termite mounds that had been broken previously by a spiny anteater. At a Queensland nature reserve where he is the ranger, he writes he has seen some astonishing sights. "For instance (a spiny anteater) will peel the bark out of logs to get at the termite colonies underneath. It does so by using its claws, which it inserts in cracks in the bark and then pulls it outward until large strips are removed. I've had to use a crowbar to peel bark from logs which the (spiny anteater) has peeled with its claws." Incidentally, the spiny anteater seems to suffer no ill effects from its ant diet, though many of the ants it eats have painful stings. Presumably, it has acquired some immunity.

Not long ago an Australian newspaper carried the headline: "Animal 'Escapologist' Gets Out of It Again." The article said that an agile and determined spiny anteater had added further luster to the records of one of Australia's most celebrated families of noted "escapologists." It went on to relate that a naturalist had put an anteater into an empty kerosene can, which he had covered with a sheet of iron. The can with its iron lid was put inside a 12-gallon pot normally used for boiling the family linen, and the lid of the pot was weighed down with bricks. During the night the anteater blithely heaved the whole thing away and headed back to the bush.

Another Australian zoologist thought he had locked a spiny anteater safely in his kitchen overnight. In the morning it was gone. Moreover, in nosing around for an escape, it had moved everything toward the center of the room, including a table, a heavy dresser and three heavy boxes! Only a fixed gas stove had resisted its efforts. On another occasion a naturalist locked a spiny anteater in a room of a suburban house in Sydney. The animal succeeded in moving a heavy upright piano out from the wall. Placed in zoo enclosures, only concrete resists the stout claws as the animal attempts to dig its way to freedom. When attacked by dogs or dingoes (wild dogs), the spiny anteater raises its sharp quills and excavates its way swiftly into the ground. If the ground is very hard, it takes a strong grip on it with its powerful claws and holds fast. The dingoes find it almost impossible to dislodge and usually retreat with bloody noses, stuck like pincushions with the quills of the creature. (A naturalist who wanted to capture a spiny anteater had to use a crowbar to lever one from its grip on the ground.)

Spiny anteaters lay usually one egg; sometimes two are

laid. Naturalists used to think that the female moved the egg to the pouch with her paws, but the strong-nailed claws are not very suitable for this task and it is more probable that the anteater's supple body allows it to lay the egg directly into the pouch.

No one knows how long it takes a spiny-anteater egg to hatch. The newly hatched young is naked, blind, helpless and about half an inch long. When it is about seven inches long the mother casts it from the pouch. It is still quite naked, with only a few short spines that are soft and offer it very little protection. The mother continues to feed it until its spines have stiffened and it can fend for itself.

As we observed earlier, the platypus hasn't a pouch; the animal's semiaquatic habits make a pouch most unsuitable for the protection and transport of young. The female platypus has acquired the practice of making burrows in which to incubate her eggs. A mated pair of platypuses inhabit a simple kind of burrow in the river bank when they are not breeding; usually a semicircular excavation between the roots of large trees. The more complicated breeding burrow is excavated by the female alone. A low, arched tunnel, from 15 to 60 feet in length, with many dead-end branches, leads to the breeding chamber. The entrances of both burrows are usually several feet above the waterline. A notable feature of all tunnels is that they conform closely to the shape of the animal's body. The tight-fitting tunnel acts as a "wringer," squeezing out the water, so that the platypus emerges comparatively dry into its living or breeding chambers. When building "platypusaries" in zoos and sanctuaries it is essential to ensure that man-made tunnels are tight-fitting. The female lines the nesting burrow with eucalypt leaves, grass, willow wands, or reeds, which she gathers mainly from the water and carries underground in the curved grip of her tail. About 15 days after mating, she retires to the burrow and plugs it at various places behind her with barriers of mud, from six to eight inches thick, which she tamps down, using her tail as a trowel. The exact purpose of these barriers isn't known. They could help to preserve heat; they could help to resist flooding, although in most instances the breeding chamber is well above flood level; or they could provide some protection against predators, such as pythons and monitor lizards.

The female platypus lays usually two soft-shelled eggs about the size of sparrow eggs. While in the oviduct, the eggs absorb nutriment and increase in volume. They are quite leathery—you can squeeze them between your fingers. Incubation lasts from 10 to 14 days, during which time the mother curls her body around the eggs. The newly hatched tiny platypuses are naked and blind, and they have baby teeth, which they later lose. (Neither the platypus nor the spiny anteater has teeth; they grind their food between horny serrations on the back of their tongues and hard ridges on their palates.)

During the first few weeks of life, the baby platypuses stay close to their mother, lapping up the milk from the teatless mammary glands in her abdomen. During this time the mother does not leave the nesting burrow. Hence, not a great deal is known about the infancy of tiny duckbills. They have been bred only once in captivity—by the Australian naturalist David Fleay, at Healesville in Victoria. His feat in 1943 made world headlines.

The tiny platypuses are blind for about eleven weeks. The young stay in the nest for a further six weeks and emerge for their first swim at about four months. About this time the mother begins to wean her babies, but they take very little food at first.

There's something eerie and awe-inspiring in watching platypuses feeding at dusk, as I have found. It's as though you've moved back into the dawn of time. Although they only use their front legs for swimming, they are swift and graceful swimmers. Diving, thumping their flat tails with pistollike smacks, they sift their food of small crustaceans, worms, aquatic insects and their larvae, and mollusks with the supersensitive nerves in their snouts. They use their snouts for nuzzling, much as ducks do. Although it forages for most of its food in the water, the platypus spends only a few hours there, at dawn and sunset. The rest of the time it spends in its burrow.

On land the platypus is clumsy, wriggling forward much like a reptile. It folds its web toes inward and "walks" on its "knuckles."

All platypuses are large eaters; adults eat about half their weight in earthworms every day, and the appetite of a nursing mother almost doubles. David Fleay found that a platypus he called Jill, who weighed only two pounds, ate one and three-quarters pounds of worms, grubs and small crayfish in a single night when she was suckling her baby.

Thus, finding sufficient food is always a problem for any zoo or sanctuary that tries to keep platypuses in captivity. Sometimes the situation becomes so desperate that the help of schoolchildren is invited in the newspapers to save a platypus from short rations. David Fleay was delighted when he found some giant earthworms, up to four feet long, near his present sanctuary in Queensland. They helped him get through one feeding crisis with his platypuses.

A platypus called Splash lived in captivity in the Sir Colin MacKenzie Sanctuary at Healesville, Victoria, for four years, and ate during that time three-quarters of a ton of worms, 200 dozen fowls' eggs and thousands of tadpoles. Another male platypus, Jack, ate three pounds of earthworms and ten yolks of hens' eggs a day. (Perhaps it was because of his ample diet that Jack lived 17 years in captivity.)

Initially it was very difficult to keep platypuses at all in captivity—let alone send them abroad to zoos. A number died on the way to New York. Then in 1922 the Bronx Zoo was able to display "the most wonderful of all living mammals" for 49 days—after which the creature died. Thanks to the work of David Fleay, the Bronx Zoo was able later to keep two platypuses, Cecil and Penelope, in good health for over ten years. Penelope escaped from the platypusary in 1957 and was never found. Cecil died two years later.

Platypus and spiny-anteater males both have spurs on the back of each hind leg, which they use to hold the female during mating. The spur of the platypus is connected to a poisonous sac. No deaths from platypus poisoning are recorded, but trout fishermen have been painfully stung when trying to remove a fly hook taken by a platypus. (I've seen a platypus strike at a fly when I have been fishing.)

In an attempt to unlock the secrets of monotremes, scientists are fitting them with tiny radio transmitters. They can locate them even when they are underground and can take their temperatures when they can't see them. At Monash University, Melbourne, Dr. E. H. M. Ealey was able to track a spiny anteater for nearly two months and take its temperature twice a day although he only saw the animal seven times in that whole period. Both the platypus and the spiny anteater have inadequate temperature control. Spiny anteaters cannot sweat or pant to control their body temperature. In arid regions they have to burrow into the earth on very hot days; in colder regions the spiny anteater hibernates. About this Dr. Ealey says: "We have found, for instance, that a spiny anteater's heart rate may fall as low as two beats per minute when it is cold. . . . Our results so far indicate that the spiny anteater is a true mammalian hibernator. It does not merely become torpid with cold like a reptile."

Both unique creatures are rigidly protected in Australia and should survive, provided their habitats are not unduly interfered with. In the past a fur trade menaced the platypus. (Fifty to seventy skins with their thick, smooth fur make a splendid rug.) The spiny anteater seems to be a match in most instances for dingoes, wild dogs and foxes, though these predators do eat some. Others are run over by cars. Platypuses appear to have very few natural enemies, but pythons, large monitor lizards and water rats sometimes take them, and so do crocodiles in Australian tropical waters. The introduced European rabbit possibly poses the greatest threat. The rabbits build their burrows in the banks of rivers where the platypuses breed. Their disappearance from many settled areas is attributed to rabbits, for without suitable soil for breeding burrows they cannot survive.

We have come to the end of the wondrous chronicle of eggs. It began with the primitive eggs of the lancelets, descendants of the first vertebrates. Then the marvelous reptilian egg with its internal pond made the conquest of the land possible and led to the development of birds, and, ultimately, to the monotremes, primitive mammals that laid eggs but suckled their young. From here to more advanced mammals and man is, in terms of evolution, but a short step. Advanced mammals, of course, don't lay eggs but instead nourish their young in their wombs. All life is a continuous whole, vast, interlinked, interwoven. I have a zoologist friend who says he's always tempted to lift his hat whenever he sees a starfish. It's a private joke, but I know what he means. The ancestors of the first vertebrates probably evolved from a distant relative of the starfish!

INDEX

359